Scientific Data Analysis

An Introduction to Overdetermined Systems

Richard L. Branham, Jr.

Scientific Data Analysis

An Introduction to
Overdetermined Systems

With 36 Illustrations

Springer-Verlag
New York Berlin Heidelberg
London Paris Tokyo Hong Kong

Richard L. Branham, Jr.
Jefe Area de Matematicas
  y Del Centro de Computos
Centro Regional de Investigaciones
  Científicas y Tecnológicas
Casilla de Correo 131
5500 Mendoza
Argentina

Library of Congress Cataloging-in-Publication Data
Branham, Richard L.
    Scientific data analysis: an introduction to overdetermined
systems/Richard L. Branham, Jr.
        p. cm.
      1. Mathematical statistics—Data processing.   2. Equations—
Numerical solutions—Data processing.   I. Title.   II. Title:
Overdetermined systems.
QA276.4.B724   1990
519.5'0285—dc20   89-28533

Printed on acid-free paper.
© 1990 by Springer-Verlag New York Inc.
Softcover reprint of the hardcover 1st edition 1990

Typeset by Asco Trade Typesetting Ltd, Hong Kong.

9 8 7 6 5 4 3 2 1

ISBN-13:978-1-4612-7981-5   e-ISBN-13:978-1-4612-3362-6
DOI: 10.1007/978-1-4612-3362-6

Preface

This monograph is concerned with overdetermined systems, inconsistent systems with more equations than unknowns, in scientific data reduction. It is not a text on statistics, numerical methods, or matrix computations, although elements of all three, especially the latter, enter into the discussion. The reader I have in mind is a scientist or engineer who has gathered data that he or she wants to model by a mathematical system, perhaps linear, perhaps nonlinear, and solve to obtain the best estimates, in some sense of the term "best," of various parameters.

Because the calculations will be performed on a digital computer, the first chapter discusses floating-point numbers and their effect on mathematical operations. The chapter ends with some methods for accurately summing floating-point numbers, an operation frequently required in numerical work and one often done by the worst possible method, recursive summation.

Chapter 2 gives a brief review of linear algebra and includes vector and matrix norms and condition numbers of matrices and linear systems.

Chapter 3 presents some ideas for manipulating sparse matrices. Frequently, time or memory can be saved by use of sparse matrix techniques. The subject is extensive and the chapter is only indicative of the many techniques available. Although Chapter 3 is somewhat extraneous to the rest of the book, Chapter 5, on linear least squares, makes use of the compressed storage mode for the symmetric matrices discussed in Chapter 3.

Chapter 4 introduces overdetermined systems and gives a brief history of the subject. The chapter contains criteria for the rejection of discordant observations, essential for achieving a good solution when one employs the method of least squares.

Chapter 5, the longest in the book, deals with linear least squares. The least squares criterion is by far the most widely used criterion for solving over-determined systems. Discussed are solutions by the formation of normal equations and their Cholesky decomposition, both classical and square root-free, and the derivation of the variance–covariance and correlation matrices. Solutions by orthogonal transformations, Givens and Householder, come next. Although orthogonal transformations are popular among many, I feel that their virtues, while real, have been exaggerated and that for most problems the normal equations are adequate and have virtues of their own, such as giving a faster solution and needing less computer memory. The chapter includes iteratively reweighted least squares, an alternative to outlier rejection for avoiding the nefarious effects of discordant observations, and least squares with linear equality constraints.

The L_1 method, useful when discordant data are present, comes next in Chapter 6. Because the most widely used algorithm for an L_1 solution is a modification of Dantzing's simplex method for linear programming, the chapter gives a brief summary of the concepts of linear programming. Also discussed is how the L_1 criterion avoids being derailed by discordant observations, so damaging, unless eliminated, when one uses least squares.

A number of nonlinear techniques, both relying on the gradient and gradient-free, are presented in Chapter 7. Some of the techniques, like the Levenberg–Marquardt method, are restricted to the least squares criterion, and others, like Nelder and Mead's simplex method, can be used with any criterion.

The final chapter discusses the singular value decomposition (SVD), a technique of genuine utility, as with a total least squares solution, but one that can easily be abused, as Section 8.4 shows. As with orthogonal transformations the virtues of the SVD have been exaggerated and even when used correctly the SVD has competition. Total least squares, for example, can also be calculated by normal equations (see Section 8.3). The SVD remains, nevertheless, a powerful tool that should be in every scientist's mathematical toolbox.

Programs and subroutines, most in FORTRAN, a few in BASIC, and one in C, illustrate many of the techniques presented. Despite being the default and most widely used scientific programming language, FORTRAN suffers certain deficiencies. The example of a linked list representation of a sparse matrix had to be given in C because FORTRAN does not support dynamic data structures. The proposed FORTRAN-8x revision remedies many of the language's deficiencies, but one cannot help wondering why it took the scientific programming community so long to appreciate features that a farsighted language like PL/I implemented over twenty years ago and that C incorporates now. But, for better or worse, FORTRAN remains the standard, and to have given most of the programming examples in PL/I, as I first thought of doing, would probably have consigned the book to limbo. One might get away with it in C, but even with that language I have my doubts.

For the reader's convenience the illustrative programs are collected at the end of the chapters, before the References, and are presented as figures. A few routines in BASIC, not complete programs or subroutines, are interspersed throughout the text. I have used these programs and subroutines in various research projects and they seem to function correctly. But I make no guarantees that they are error-free, and the reader should use them with this proviso in mind.

These programs and subroutines were developed on the VAX-11/780 (VAX and VMS are trademarks of the Digital Equipment Corporation) of the Centro Regional de Investigaciones Científicas y Tecnológicas (CRICYT) with VMS 3.5. In its original configuration the VAX incorporated 1 MB of physical memory, subsequently expanded to 3 MB and then 4 MB. Anyone who desires the programs in machine-readable form can send me a tape, and I will write the programs on it in either 800 b.p.i. or 1,600 b.p.i. (not 6,250 b.p.i.) and in either ASCII or EBCDIC. (The ideal, of course, would be to make the software available by electronic mail, but unfortunately at the time of writing CRICYT is still not connected to an international network such as BITNET.) The routines in BASIC were coded on a SHARP EL-5500 II pocket calculator/ computer with 4 K of memory.

This book grew out of experiences I have had with astronomical data reduction, particularly the analysis of astrometric observations, both radio and optical, when I worked at the U.S. Naval Observatory (Washington, D.C.) and from various courses given at CRICYT, courses on programming, computational linear algebra, and computational techniques with sparse matrices, and a course on computational linear algebra at the Argentine National University of La Plata.

It is my pleasant duty to thank the personnel of CRICYT's Comunicación Visual (S. Farías de Candia, D. Dueñas, H. Miranda, and D. Rosales) for preparing the illustrations. Special thanks are due to my secretary, Mrs. Beatriz Grillo, for her careful typing of the manuscript, and in a language that is not her native tongue. (English is my native language, and any defects in style or usage are directly attributable to me.) She also composed the programming examples. (Because this was done on an IMB Composer the term "composed" is appropriate.) I would like to thank my wife, Rosa, for first suggesting that I work on a book of this nature and also for her, and my daughter Maria's, patience when I would absent myself to an obscure corner of the house to write—with a classical music background (Bruckner, Mahler, and Verdi were favorites).

Mendoza   *Richard L. Branham, Jr.*
The Argentine

Contents

CHAPTER 1

Properties of Floating-Point Numbers

1.1. Introduction

Although the topic of this book is the solution by computer of overdetermined systems, it is advisable to begin with a discussion of floating-point numbers. Aside from occasional projects of a special nature, scientific computations are carried out with floating-point numbers. In the early days of computing, when floating-point hardware was not available, fixed-point was used, and we can still find early papers in the literature that are based on fixed-point calculations. But the advantages of floating-point are so manifest that the hardware to implement it was soon made or floating-point operations simulated; simulation is still the rule with microcomputers, although floating-point coprocessors are becoming more common. If, then, our algorithms are executed with floating-point operations, a measure of prudence dictates that we be familiar with the workings of floating-point.

1.2. Representation of Floating-Point Numbers

Floating-point numbers are characterized by four parameters: the base, B, of the number system; the precision, p, of the representation; the lower limit, L, and the upper limit, U, of the representation. Common bases are binary, octal (little used now), decimal, and hexadecimal. The representation of a floating-point number F is

$$F = \pm \cdot \left(\frac{d_1}{B} + \frac{d_2}{B^2} + \cdots + \frac{d_p}{B^p} \right) B^e, \tag{1.1}$$

where d_i is a digit of the base, $0 \le d_i \le B - 1$, and e is an exponent in the range $L \le e \le U$. The portion of the number within parentheses is the fraction or mantissa. A normalized number has the first digit, d_1, distinct from zero; otherwise the number is denormalized.

Normalization, by preserving all of the precision of the number, is desirable. Leading zero digits are mere place holders and crowd out significant digits on the right. Unfortunately, operations with floating-point numbers often result in their denormalization. Suppose we add two numbers with different exponents. The two exponents are brought into agreement by shifting the digits of the number with the lower exponent to the right. Significant digits are dropped on the right—"fall into the bit bucket" is the expression commonly used—and replaced by place-holding zeros on the left. With exponents of vastly different magnitude all of the significant digits of the smaller number will be lost. The addition of 10^{30} to 10^{-30} on most computers gives 10^{30}; all of the digits of 10^{-30} are shifted into the bit bucket. In a series of calculations the precision loss caused by the unavoidable denormalizations brings about an erosion, usually gradual but sometimes spectacularly abrupt, in the significant digits of the result.

Let us consider some concrete examples of floating-point representation. The VAX series of computers, manufactured by the Digital Equipment Corporation (DEC), is popular for scientific computing. A floating-point number on a VAX (a byte addressable computer), occupies four bytes in single-precision, as shown in Figure 1.1. The bits in the number run right to left from zero to thirty-one. The eight bits from seven to fourteen are the exponent. Bits zero to six are the most significant bits of the fraction and bits sixteen to thirty-one the least significant bits. Bit fifteen is the sign bit of the number. If it is zero the number is positive; if it is one the number is negative. This way of taking care of the number's sign, the way we are most accustomed to, is referred to as sign and magnitude.

A different scheme is used for the eight bit exponent. These bits correspond to numbers from zero to $2^0 + 2^1 + \cdots + 2^7 = 255$. To avoid an explicit sign bit the exponent is biased by 128: the bias quantity 128 is substracted from the exponent to obtain the real exponent. Zero is $0 - 128 = -128$ and 255 is $255 - 128 = 127$ for a range of -128 to 127. But the VAX reserves the exponent -128 exclusively for zero regardless of the mantissa. Sign and magnitude and biasing are two ways of handling the sign problem. A third way, complementing, will not be discussed as it is irrelevant for the present development of our ideas. The interested reader may refer to, among other references, Section 12.2 of Sterbenz's (1974) excellent book on floating-point computation.

It appears that the VAX uses twenty-three bits for the mantissa, seven for the most significant bits, and sixteen for the least significant. But as the numbers are normalized the first bit is always one and is not represented explicitly. With the implicit first bit there is a total of twenty-four bits. We

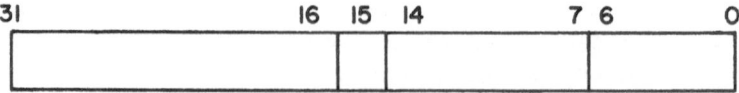

Figure 1.1. Single-precision VAX floating-point number.

Figure 1.2. Single-precision IBM 370 floating-point number.

can represent numbers varying in range from $0.1_2 \cdot 2^{-127} = 2.9 \cdot 10^{-39}$ to $0.111 \ldots 1_2 \cdot 2^{127} = 1.7 \cdot 10^{38}$. (Remember that the VAX keeps floating-point numbers normalized. The smallest mantissa, therefore, is 0.1, base two, usually written 0.1_2 to show the base explicitly.)

Take another example. The IBM 370, whose architecture has been continued with little modification on the 4300 series, is also a byte addressable computer. Single-precision floating-point numbers occupy, like the VAX, four bytes, as shown in Figure 1.2. Unlike the VAX the bits are numbered left to right. Bit zero is the sign bit, zero for positive numbers and one for negative numbers. Bits one to seven are the exponent, biased sixty-four: subtract sixty-four from the exponent to obtain the real exponent. A total of twenty-four bits, eight to thirty-one, contain the mantissa. The base of the representation is sixteen, or hexadecimal. Whereas the VAX has a precision of twenty-four binary digits, the IBM 370 has a precision of six hexadecimal digits.

Although sixteen is a power of two, a hexadecimal computer is not a binary computer. A normalized hexadecimal number with 1_{16} as the first digit begins with the bits 0001, but if the first digit is F_{16} the bits are 1111. We lose between zero and three bits of the twenty-four. The average number of bits will be less than twenty-four; we have less precision in the mantissa compared with a binary machine.

The range of single-precision floating-point numbers with the IBM 370 varies from $0.1_{16} \cdot 16^{-64} = 5.4 \cdot 10^{-79}$ to $0.FFFFFF_{16} \cdot 16^{63} = 7.2 \cdot 10^{75}$. With one less bit in the exponent, the range is nevertheless greater than the VAX's. We suspect that something is amiss, that something must be lost to compensate for the greater range. Correct. Resolution is lost. The IBM exponents are powers of sixteen, the VAX exponents powers of two. For a given power m we have $16^m = (2^4)^m = 2^{4m}$. There are four VAX exponents for every IBM exponent, a finer resolution for the VAX. For example,

$$16^{-31} = 2^{-124}$$

$$\left.\begin{array}{c} 2^{-123} \\ 2^{-122} \\ 2^{-121} \end{array}\right\} \quad \text{no IBM representation}$$

$$16^{-30} = 2^{-120}.$$

Both the VAX and the IBM 370 are byte addressable computers. Not all computers are. The DEC System 20, for example, uses a thirty-six bit word as the addressable entity. Other computers, such as the obsolete IBM 7090, do not use biased exponents but rather exponents with sign and magnitude. It is obviously not possible to give examples of all the combinations. The VAX and IBM, being popular computers, are at least good representative examples.

1.3. Characteristics of Floating-Point Numbers

There is not an infinity of floating-point numbers. The finite precision and the upper and lower limits for the exponents assure this. How many numbers are there? There are $(U - L + 1)$ exponents. For a given exponent and base there are p digits for each mantissa, allowing for B^p possibilities. But we assume that our numbers are normalized $(d_1 \neq 0)$ and must eliminate all of the mantissas with $d_1 = 0$. This leaves $B^p - B^{p-1}$ possibilities. The numbers may be positive or negative. We arrive at the formula

$$2(B^p - B^{p-1})(U - L + 1) + 1 = 2(B - 1)B^{p-1}(U - L + 1) + 1, \quad (1.2)$$

the final 1 coming from the unique zero for the system.

The VAX, with $B = 2$, $p = 24$, $U = 127$, $L = -127$—remember that all numbers with exponent -128 are considered zero—has 4,278,190,081 in its single-precision floating-point system. The IBM 370, with $B = 16$, $p = 6$, $U = 63$, $L = -64$, has 4,026,531,841 numbers. We see once again that the VAX's systems must have greater resolution. With a smaller range of exponents it can nevertheless represent more numbers.

Floating-point numbers are not evenly distributed through their range, only through successive powers of the base. Consider the decimal system with $p = 1$, $U = 1$, $L = -1$. Numbers with exponent 10^{-1} are separated by $0.1 \cdot 10^{-1} = 0.01$, with exponent 10^0 by $0.1 \cdot 10^0 = 0.1$, and with exponent 10^1 by $0.1 \cdot 10^1 = 1.0$.

Because of its finiteness not any given real number can be represented on our floating-point system. Real numbers greater than U and less than L are not represented at all. If we attempt arithmetic with a number like 10^{80} in single-precision on a VAX the program comes to an abrupt stop, emitting an error message like "exponent overflow". Even numbers within the exponent range rarely have an exact representation on the floating-point system. Let x be the given real number and $fl(x)$ its approximation in our system. The approximation can be made in one of two ways. With rounding, the number $fl(x)_R$ in our floating-point system nearest the given number x is taken. With chopping, sometimes called truncation, the number $fl(x)_c$ closest to the smaller magnitude is used.

Assume that the number x lies in the interval $[B^{e-1}, B^e]$, where e is some exponent between U and L. The floating-point numbers are evenly spaced with a separation of B^{e-p}. Therefore, the closest number $fl(x)_R$ must be within a distance $\frac{1}{2}B^{e-p}$ of x. We have for the absolute error

$$|fl(x)_R - x| \leq \tfrac{1}{2}B^{e-p}. \tag{1.3}$$

Because $B^{e-1} \le x$, the relative error is

$$\frac{|\text{fl}(x)_R - x|}{|x|} \le \tfrac{1}{2}B^{1-p}. \tag{1.4}$$

With chopping the closest number $\text{fl}(x)_c$ can be as far away as B^{e-p}. For the absolute error

$$|\text{fl}(x)_c - x| < B^{e-p}. \tag{1.5}$$

Notice the inequality. There is no possibility of an equality as with rounding. It is quite possible that our real number x lies exactly in the middle of two floating-point approximations, and the absolute rounded error is $\tfrac{1}{2}B^{e-p}$. But with chopping we never reach the maximum error B^{e-p} because then there would be a number in $\text{fl}(x)_c$ that exactly corresponds to x, and the absolute error would be zero. For the relative chopping error we obtain

$$\frac{|\text{fl}(x)_c - x|}{|x|} < B^{1-p}. \tag{1.6}$$

With Eq. (1.4) we calculate that a VAX has a relative rounded error of $5.96 \cdot 10^{-8}$. From Eq. (1.6) the relative chopping error for an IBM 370 is $9.54 \cdot 10^{-7}$. Iterating what was said previously about the VAX's greater resolution of floating-point numbers, we can say that not only is the VAX's resolution better, but that its relative error is sixteen times better than the IBM's.

Rounding is more desirable than chopping. Not only is the relative error smaller with the former, it is also not systematic: sometimes we round up to the next higher floating-point approximation and sometimes down to the next lower one. With chopping, our approximate number is always less in absolute value than the given real number. The result is a greater accumulation of error after a series of chopping operations compared with rounding operations. Following the discussion in Section 3.12 of Sterbenz's (1974) book and also Exercises 16 and 17 of Chapter 3 of that book, we can derive formulas for the error growth in both rounding and chopping arithmetic.

We are summing a series of N numbers, each one of which has an error ε_i. The numbers are normalized to integers so that the ε_i are the errors that arise from rounding to the nearest integer, or $-\tfrac{1}{2} \le \varepsilon_i \le \tfrac{1}{2}$. We assume that the ε_i are independent random variables drawn from a uniform distribution between $-\tfrac{1}{2}$ and $\tfrac{1}{2}$. The error in a sum of N numbers is

$$\varepsilon = \sum_{i=1}^{N} \varepsilon_i. \tag{1.7}$$

The distribution function of ε has a mean of zero and a variance of

$$\varepsilon = N \int_{-1/2}^{1/2} \varepsilon^2 \, d\varepsilon = \frac{N}{12}. \tag{1.8}$$

Therefore, the standard deviation of the distribution of the errors is $\sqrt{N/12}$. With chopping arithmetic, ε_i is distributed as $0 \le \varepsilon_i < 1$. After a series of N

summations the expected value of ε is $N/2$, not zero, with a standard deviation of $\sqrt{N/12}$. For large N, the $N/2$ factor dominates.

Combining these two results we see that with rounding arithmetic the growth of error is $O(\sqrt{N})$, and with chopping arithmetic, $O(N)$. Forsythe and Moler (1967, p. 90) arrive at a similar conclusion using a heuristic argument. They also discuss more completely, in Chapter 20, the errors in floating-point operations.

Let us consider an example of the error growth in floating-point operations. The example is standard, given by Sterbenz (1974, Sec. 4.1) and Scheid (1968, p. 114) among others. Analytically,

$$\int_0^{\pi/2} \sin x \, dx = 1. \tag{1.9}$$

If we were unaware of this analytic result—admittedly this is farfetched—then we could integrate numerically with Simpson's rule. If h is the step size, then between the limits x_1 and x_N Simpon's rule is

$$\int_{x^1}^{x_N} f(x) \, dx = \frac{h}{3}(f_1 + 4f_2 + 2f_3 + \cdots + 2f_{N-2} + 4f_{N-1} + f_N). \tag{1.10}$$

Books on numerical methods prove that the error in Simpson's rule is $O(h^5 f^{iv})$, where f^{iv} is the fourth derivative of the function f evaluated at some unknown point in the interval $[X_1, X_N]$. Obviously, as we make h smaller, the error also becomes smaller. With this consideration in mind suppose that we go hog-wild and divide the interval $0 - \pi/2$ into $2^{16} = 65{,}536$ subintervals, for a step size of $h = \pi/(2 \cdot 2^{16}) = 2.3968 \cdot 10^{-5}$. Equation (1.10) may be implemented in a straightforward manner with any programming language. Sterbenz gives a short FORTRAN program, but notice that rather than entering $\pi/2$ as 1.570796 it is preferable for us to calculate it as $2.0 * \text{ATAN}(1.0)$. This is slightly more time-consuming, albeit only trivially so, but there is less chance of error.

If our simple program is run on an IBM 370—the author actually ran it on an IBM 4341, which has nearly the same architecture—the result is 0.99309123. Running it on a VAX-11/780 gives 0.9998720. Aside from the shock of seeing one result so different from another it is evident that rounding produces a result correct to four digits, whereas chopping produces only two digits. The shock of seeing that a fine subdivision of the interval yields mediocre results dissipates upon our realizing that the interval is too fine, and results in a summation of many terms with the accompanying erosion of significant digits. With a sum of 65,536 terms, each one with a VAX error of $5.96 \cdot 10^{-8}$, and the error growth proportional to the square root of the number of terms, the final error should be of the order $1.5 \cdot 10^{-5}$, exactly what is observed. With an IBM 370 error of $9.54 \cdot 10^{-7}$ and with linear growth of the accumulated error, we expect a final error of about $6.2 \cdot 10^{-2}$. The IBM 370 actually does better than this and gives two, rather than one, significant digits in the result. Nevertheless, the difference in precision between chopping and rounding is startling. The moral is obvious: rounding machines are superior to chopping machines for numerical work.

A quantity of interest with floating-point arithmetic is the machine epsilon, the smallest number ε that admits the inequality $1 + \varepsilon > 1$. Its importance arises in the control of iterative loops. In many algorithms we correct an approximate value until the difference of the current value and the previous value is smaller than some preassigned number, say δ. Considering the current value V_n and an old value V_0, we want the relative error smaller than δ

$$\left| \frac{V_n - V_0}{V_0} \right| \leq \delta.$$

A little algebra gives

$$|V_n| < |V_0|(1 + \delta).$$

If the specified tolerance δ is smaller than the machine epsilon this requirement will never be met. With single-precision on a VAX, for example, if we specify $\delta = 10^{-10}$, $1 + \delta = 1$ and the inequality can never be satisfied. The new value will equal the old value, and the program will fall into an infinite loop.

An approximation to the machine epsilon is easily calculated. The following FORTRAN segment can be incorporated into our programs to calculate the machine epislon at run time:

```
EPS = 1.0
EPSPL1 = 1.0E38
DO WHILE (EPSPL1.GT.1.0)
    EPS = EPS/1.1
    EPSPL1 = 1.0 + EPS
END DO
```

For greater accuracy we could replace the 1.1 by 1.01, or even 1.001, but as the machine epsilon is not needed to great accuracy this refinement is unnecessary.

The discussion so far has been concerned with single-precision. Most computers also incorporate double-precision and, in some cases, extended precision. Double-precision adds more bits to the mantissa to achieve greater precision in the representation. Byte addressable computers typically assign eight bytes to double-precision numbers. Both the VAX and the IBM 370 use fifty-six bits (one of them implicit on the VAX) for double-precision numbers. The term "double-precision" is somewhat of a misnomer. It is true that twice the bytes are used, but the mantissa receives more than double the number of bits. The exponent retains its eight bits. Sometimes, however, more bits may be given to the exponent. Because the VAX's exponent range is small, it incorporates an optional alternative double-precision, G-FLOATING. This takes three bits away from the mantissa and gives them to the exponent. The eleven bit exponent, biased 1024, runs from -1023 to 1023 (-1024 is reserved for zero), and increases the range to $5.6 \cdot 10^{-309}$ through $9.0 \cdot 10^{307}$. The three bits lost in the mantissa, corresponding to one decimal digit, are little missed. Extended precision is an option on many machines. With a VAX, for example, extended precision occupies sixteen bytes, allocating fifteen bits to the exponent and 113, one bit implicit, for the mantissa. Numbers in the range $8.4 \cdot 10^{-4933}$ to $5.9 \cdot 10^{4931}$ can be represented. On a VAX single-precision gives about seven decimal digits of representation, double-precision about seventeen, sixteen for G-FLOATING, and thirty-four for extended precision.

There are algorithms that reveal the properties of the computer's floating-point arithmetic. Figure 1.4 gives a short FORTRAN program that combines Malcolm's algorithm (1972) with improvements that Gentleman and Marovich (1974) suggest, and also incorporates a test for an add/subtract guard digit. This test is taken from a PASCAL program published in the February 1985 *BYTE* (Karpinski, 1985). The interested reader may refer to these publications for an explanation of how the program works. He should also be warned that the program is not infallible. A computer with weird arithmetic properties can fool it. But the program works correctly on most computers and offers a way of determining at run time, if incorporated into a larger program, the properties of the computer's arithmetic.

1.4. Violation of the Laws of Arithmetic

Most of us learned in high school or college certain laws that govern arithmetic operations. These go by names such as the associative law, the distributive law, and others. Because they are laws they must be correct, right? Absolutely wrong. In the ideal world of pure mathematics, where numbers range from minus infinity to infinity and word length does not exist, the laws indeed are valid. But in the compromised world of finite machines we encounter some unpleasant facts. The finite word length of floating-point numbers and the upper and lower exponent limits invalidate the applicability of the arithmetic laws and result in calculations whose precision is compromised. We have already talked about the erosion of significant digits in a numerical integration.

Sterbenz (1974) provides a rigorous analysis of the breakdown of the arithmetic laws. Many readers may find it equally illuminating—or even more illuminating if they have an aversion to abstract mathematics—to see concrete examples of the laws' breakdown and why it happens. In any event that is what we shall provide. The mathematically more sophisticated reader will undoubtedly prefer Sterbenz's discussion.

The commutative laws,

$$a + b = b + a,$$
$$ab = ba,$$

(1.11)

are obeyed, but we run into difficulties with the associative laws:

$$(a + b) + c = a + (b + c),$$
$$(ab)c = a(bc).$$

(1.12)

Consider the multiplicative associative law with $a = b = 10^{38}$ and $c = 10^{-38}$. On a VAX with single-precision the product ab exceeds the upper limit for exponents; the program comes to a screeching halt. But the product bc is one and when multiplied by a we have a result of 10^{38}.

As a less drastic example, take the additive associative law with $a = 10^{-15}$, $b = 1$, and $c = -1$. If, once again, we use a VAX and single-precision, $a + b = 1$ because a is less than the machine epsilon. Therefore, $(a + b) + c = 1 - 1 = 0$. But $b + c = 0$ and upon summing a we obtain 10^{-15}, $a + (b + c) = 10^{-15}$.

The distributive law,

$$a(b + c) = ab + ac, \tag{1.13}$$

is not obeyed. It is easy to find some b's and c's for which the sum $(b + c)$ overflows. Other choices have less drastic but still nefarious consequences. Use single-precision on a VAX and take $a = 1.5, b = 1.5 + 2^{-22}$, and $c = 2^{-22}$. We find that $ab + ac = 2.2500010$, but $a(b + c) = 2.2500007$. What happened? Of the two results 2.2500007 is the more accurate. An examination of the individual bits of each result shows that they differ by one bit. (In FORTRAN the individual bits may be examined if we declare one variable as an array of bytes, equivalencing it to a floating variable, and printing out the separate bytes: for example,

```
        REAL * 4        A
        BYTE            AA (4)
        EQUIVALENCE  (A, AA(1))
            ⋮
        WRITE (*, 5)    (AA(I), I = 4, 1, −1)
5       FORMAT (4Z2.2)                        .
```

C allows us to declare bit fields and examine individual bits, obviating the need for the devious stratagems required in FORTRAN.) What caused the one bit difference? The disparity in magnitude between ab and ac. To align the binary points a significant bit of the smaller number, ac, is shifted by twenty-three binary places and falls into the bit bucket. The lost bit, obviously, is not restored when ab is added to ac. b is also larger than c, but in the addition $b + c$ the smaller number, c, must have its binary point aligned by a shift of twenty-two binary places. A significant bit reaches the edge of the bit bucket, but does not fall in. The upshot is that $a(b + c)$ preserves one more significant bit than $ab + ac$.

The cancellation law fails with floating-point operations. This law states that if

$$ab = ac \quad (a \neq 0), \tag{1.14a}$$

then

$$b = c. \tag{1.14b}$$

For our practical example we will take $a = 2, b = 9$, and $c = 9 + 10^{-9}$. This time, however, instead of using a VAX we shall, for greater variety, use a Sharp EL-5500 II pocket calculator/computer. This little machine has 4 KB of memory (more than the original IBM 1401, a big computer of its day, the early 1960s), only one precision, ten decimal digits because its base is decimal, a machine epsilon of $5 \cdot 10^{-10}$, and incorporates an add/subtract guard digit. With the given a, b, c clearly $b \neq c$, but $ab = 18$ and $ac = 18$, whereas the cancellation laws leads us to expect $ab \neq ac$. The difficulty arises, once again, from losing a significant digit. When c is multiplied by a the significant one of 9.000000001 drops off the right end—we really cannot talk of falling into the bit bucket because the numbers are decimal—giving a product of eighteen, the same as ab.

Finally, the relation

$$a\left(\frac{b}{a}\right) = b \qquad (1.15)$$

is not obeyed. This is really a special case of the multiplicative associative law,

$$a(a^{-1}b) = (aa^{-1})b,$$

and the previous discussion should have convinced us that we will have troubles. Nevertheless, for the doubting Thomases, use, once again, a Sharp pocket calculator/computer and take $a = 3$, $b = 1$. Instead of one we obtain 0.999999999. The source of the difficulty is easy to see. $b/a = 1/3$, which produces an infinite $0.333\ldots$ in the ideal world of real numbers. Our computer truncates this to 0.333333333, which when multiplied by three gives the result 0.999999999.

In summary, the arithmetic laws fail because of the limits on the range of the exponents and the finite length of the floating-point number. Any real number, written in scientific notation, whose exponent exceeds U does not exist as far as the computer is concerned. Sometimes the difficulty of exponent overflow can be handled by checking for it, and setting the number to the largest one the computer can represent. This fix at least prevents a program crash, but must be treated carefully. Exponent underflow, the real number's exponent is less than L, presents less of a problem. The real number can generally be represented by zero—and most computers' hardware treats underflow by setting the number to zero—without much inconvenience. In other instances the laws fail because of finite word length. To sum two numbers their exponents must be equal. When the exponents are unequal they must be made equal by denormalizing the smaller number. This introduces nonsignificant zero bits in the high-order bit positions of the mantissa and shifts significant bits out of the number. Which bits, if any, fall into the bit bucket depends on the order in which the arithmetic operations are performed.

It *is* true that the laws are approximately obeyed. In practice, we compute with floating-point expecting, at worst, only a gradual erosion of the significant digits, and avoiding such matters as comparisons for equality between floating-point numbers. Kernighan and Plauger (1974, p. 92) state the situation well: "Floating-point numbers are like sandpiles: every time you move one, you lose a little sand and pick up a little dirt." But sometimes inattention to the vagaries of floating-point computations leads to disaster. Consider the linear system

$$
\begin{pmatrix} 1 & \frac{1}{3} & 0 \\ 2 & \frac{2}{3} & 0 \\ 0 & 0 & 1 \end{pmatrix} \begin{pmatrix} X_1 \\ X_2 \\ X_2 \end{pmatrix} = \begin{pmatrix} 1 \\ 1 \\ 1 \end{pmatrix}. \tag{1.16}
$$

Because the matrix is singular, column two is simply column one divided by three, there is no solution. Nevertheless, if we use Cramer's rule some computers will give, as a "solution" of Eq. (1.16),

$$
\begin{pmatrix} X_1 \\ X_2 \\ X_3 \end{pmatrix} = \begin{pmatrix} 3333334 \\ -10000000 \\ 1 \end{pmatrix}. \tag{1.17}
$$

The trouble here arises from $\frac{1}{3}$ and $\frac{2}{3}$ having no exact representation. The former is stored as 0.3333333 and the latter as 0.6666667. Instead of the determinant of the system's being zero, it is calculated as 10^{-7} and Cramer's rule merrily churns out the solution. Here the result is not even approximately correct; it is wrong. A contrived example such as this, of course, exists to make a point. In real life the matrix of Eq. (1.16) might not arise in a month of Sundays. The example does, however, warn us of the dangers of an overly uncritical attitude towards floating-point computations. Despite the odds being in our favor, we just might, like Napoleon at Waterloo, who estimated the odds as ninety in his favor and ten against, come to grief.

1.5. Accurate Floating-Point Summation

The failure of the arithmetic laws means that expressions such as abc and $a + b + c$ have no unique interpretation; the result depends on the order of evaluation. Summing n numbers,

$$
S = \sum_{i=1}^{n} a_i, \tag{1.18}
$$

occurs sufficiently frequently, as in forming scalar products of vectors and matrix inner products, that we want it done as accurately as possible. Because the associative law of addition is invalid with floating-point operations, we expect that some ways of forming the sum S will give better results than others.

S is generally formed by recursive summation:

$$
S_0 = 0,
$$
$$
S_i = S_{i-1} + a_i, \qquad i = 1, 2, \ldots, n. \tag{1.19}
$$
$$
S = S_n.
$$

Unfortunately, the error can grow $\propto n^2$. Rather than present a formal proof, which Linz (1970) gives, we will follow a heuristic line of reasoning that may be equally illuminating. Suppose that we are summing 1,000 numbers, each of the order ten in magnitude. After we add ten numbers the sum is of order 100. To add a new number the exponents must be aligned. The smaller number, a_i, will have its exponent increased by shifting digits to the right. A significant decimal digit is lost and replaced by a nonsignificant zero. After 100 numbers have been summed S is of the order 1,000. To add a_i two significant digits are lost. And so forth.

There are numerous ways to increase the precision of S. The simplest uses increased precision for S. If the a_i's are single-precision, S is declared double-precision. The a_i's are extended to double-precision before being added to S. Significant bits of a_i are merely shifted to the right, where there is now room to accommodate them, instead of falling into the bit bucket.

But what can we do if our computer does not incorporate double-precision—rare on a mainframe but sometimes the case with a simulated floating-point on microcomputers—or we are already working with double-precision and no extended precision is available? We may always simulate a higher precision—Linnainmaa (1981) gives details and portable PASCAL procedures—but this is bound to be slow. Not all VAX's, for example, incorporate extended precision in hardware. It is an option, but extended precision may always be simulated by a call to the Run Time Library routine LIB$ESTEMU. But the simulated extended precision is about seventy times slower than double-precision.

We may still accurately sum floating-point numbers without resorting to higher precision. An easy way, found by Kahan (1965), works on computers that normalize floating-point sums before rounding or chopping them, like the VAX. It will not work if the sum is rounded or chopped to single-precision, or double-precision if that is what we are working with, before it is normalized. Such information may be obtained by a detailed study of the machine manuals or, more simply, by testing. Testing is the only way to see what microcomputer-simulated floating-point does. Some software supports Kahan's trick, other software does not. To see how the trick works we will consider a small program segment. Although BASIC is terrible language for numerical work, it is nevertheless popular with microcomputer users. As a sop to them and also to demonstrate that even BASIC can accurately sum numbers the segment uses BASIC, one of the few examples in this book that employs that language.

Suppose that our n floating-point numbers are contained in an array. A recursive sum in BASIC is

```
10   S = 0.0
20   FOR I = 1 TO N
30   S = S + A(I)
40   NEXT I
```

BASIC coding for Kahan's trick is

```
10   S = 0.0
20   S2 = 0.0
30   FOR I = 1 TO N
40   S2 = S2 + A(I)
50   T = S + S2
60   S2 = (S − T) + S2
70   S = T
80   NEXT I
```

$S2$ is an estimate of the error when $S = T$ was last rounded or chopped. Statement 40 compensates for the error. The parentheses in statement 60 are necessary because they cause the difference $S − T$ to be evaluated first and, therefore, with little or no error because it is normalized before it is rounded or chopped.

When Kahan's trick works it is much faster than simulated higher precision. A VAX-11/780 summed 20,480 double-precision floating-point numbers in 0.77 seconds. Summing with simulated extended precision, because hardware extended precision was not available on this particular computer, gave the same sum to fifteen decimals, but the computer used 29.67 seconds.

If our computer or software does not permit Kahan's trick, we can resort to a slightly more complicated algorithm, Linz's (1970) binary summation. The difficulties of recursive summation arise from S's becoming much larger than the individual a_i's. As a palliative measure we may prevent, or a least minimize, the disparity in magnitude between S and a_i by using a series of accumulators instead of just S. Sum a_1 to a_2 to form an intermediate sum S_1, a_3 to a_4 to form S_2, and so on. Then sum S_1 to S_2 to obtain S_{11}, S_3 to S_4 to obtain S_{22}, and so forth. The process continues until all the numbers are summed. The very appropriately named binary summation can be illustrated schematically as in Figure 1.3. Linz proves that the growth of error with binary summation is proportional to $n \log_2 n$, n being the number of terms summed.

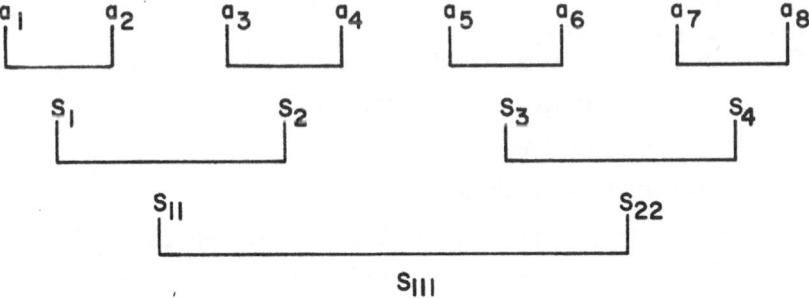

Figure 1.3. Binary summation.

It may appear as if binary summation is restricted to n's being a power of two. But Walker (1971) shows how this apparent restriction may be removed. His PL/I algorithm is easily adapted to other languages. Figure 1.5 shows a FORTRAN program to implement the summation methods discussed in this chapter. Walker's version of binary summation uses an auxiliary array T, declared of dimension 100 in the program. N must be even. M is an integer equal to or greater than $\log_2 N$. T should never be dimensioned smaller than M. B is a number distinct from any of the a_i's, and not expressible as a sum of any of them. In the program it is declared as 1.0E38, close to the upper limit for single-precision floating-point exponents on a VAX. B is machine-dependent; other values would be appropriate for other computers. A study of the program should reveal the details of the algorithm's working.

Cascaded accumulators, the last summation method we shall discuss, improves on binary summation by noting that even though S will not become much greater than an a_i, nothing prevents precision loss in summing a_i to a_{i+1} should one number be much greater than the other. It is conceivable that the partial sums S_i, S_{ii}, and so forth are of the same order of magnitude, but nevertheless a_i is 10^{30} and a_{i+1} is 10^{-30}. The algorithm for cascaded accumulators, which Wolfe (1964) first introduced, is only slightly more complicated than the algorithm for binary summation and the results are more accurate. Malcolm (1971) refined the algorithm and derived bounds for the calculated sum's absolute and relative errors. Although the basic idea of cascaded accumulators is not hard to grasp, the implementation of the algorithm depends on the programming language and computer used.

With cascaded accumulators we use a series of accumulators, each one of which holds floating-point numbers whose exponents lie within a specified range. The contents of an accumulator should never become much larger than the number being summed into the accumulator. Furthermore, if each a_i ends with zeros in the least significant part of the mantissa, the a_i's exponent may be aligned without significant bits being lost. Finally, the accumulators are summed. Specifically, the algorithm for cascaded accumulators is:

1. Given $\eta + 1$ accumulators, $\alpha_0, \alpha_1, \ldots, \alpha_n$, initialize each one to zero.
2. Decompose a_i into two, or more, parts a_{i1} and a_{i2} such that a_{i1} and a_{i2} end in nonsignificant zeros.
3. Sum each a_{i1} and a_{i2} to accumulator k, where k is defined by:

$$v = \text{range of exponent}/\eta + 1 \quad [v \text{ integer}],$$

$$vk \leq \text{exponent} \leq vk + v - 1.$$

4. Sum the accumulators in decreasing order.

Some of these steps need more detail. How, for example, can we decompose a_i in step 2? If the a_i's are single-precision they may be summed directly to double-precision accumulators. The extension of a_i to double-precision in summing will fill the least significant part of the mantissa with zeros. But what if we are already working with double-precision and extended precision is not available? The decomposition of a_i can be effected in numerous ways, all of them language-dependent. PL/I offers the useful UNSPEC function. C permits zeroing of inidividual bits. FORTRAN and BASIC programmers have to be more ingenious. One possibility among many in FORTRAN for decomposing Y into $A1 + A2$, $A1$ and $A2$ ending in zeros, is:

```
REAL * 8        A1, A2, Y
BYTE            B(8)
EQUIVALENCE     (B(1), A1)
A1 = Y
DO I = 5, 8
  B(I) = '00'X
END DO
A2 = Y − A1.
```

We could also use Boolean operators as another possibility.

For step 3 we need to determine the number's exponent. There are, again, various ways to do this. Perhaps the simplest is to note that the exponent is a power of the base. Taking the logarithm with that base of the absolute value of the number, to avoid negative logarithms, will give the exponent. If our machine is binary, we could use

$$IEXP = LOG(ABS(A1))/LOG(2D0).$$

A hexadecimal machine would use LOG(16D0). Because IEXP is integer the fractional part will be truncated.

Malcolm proves that step 4, sum the accumulators in decreasing order—rather than increasing order as we might be inclined—is important for placing a bound on the relative error of the sum. He also discusses further refinements to the algorithm. We may, for example, check for an accumulator's overflowing—exceeding the exponent limits set for it. The accumulator's contents should be summed to be next higher accumulator and the accumulator reset to zero. Unless the number of terms to be summed becomes arbitrarily large, this check is unnecessary.

Figure 1.5 also includes FORTRAN coding for cascaded accumulators. The coding is specific to a computer like the VAX that has 256 possible binary exponents. We use sixty-four accumulators ($\eta = 63$). Each accumulator holds numbers for a range of four binary exponents. The LOG function returns unbiased exponents (range -128 to 127). Because FORTRAN arrays start at one, 129 is added to the exponent so that none of the dimensions for the array ACCUM become zero or negative. LEV divides the exponent by four to determine the proper accumulator for the number. The floating-point number is decomposed by the technique already described. The coding does not check for accumulator overflow.

How do the various summation methods compare among themselves in terms of accuracy and speed? To check this the author summed 20,000 random numbers, distributed between $-10,000$ and $10,000$, in single-precision on a VAX-11/780. Table 1.1 shows the results.

Table 1.1. Sums of 20,000 random numbers.

Method	Sum	Time
Recursive, single-precision	−5,331,918.0	0ʂ14
Recursive, double-precision	−5,331,912.580173492	0.35
Kahan trick	−5,331,913.0	0.28
Binary summation	−5,331,914.0	0.43
Cascaded accumulators	−5,331,913.0	4.86

From the results of Table 1.1 we may infer that, in lieu of computing the sum in a hardware higher precision, Kahan's trick, when applicable, is the best; it is certainly the fastest. Cascaded accumulators come next. The greater accuracy in the sum compensates for the slower execution speed. The program itself is only slightly more complicated than that for binary summation. Part of the reason for the poor timing for cascaded accumulators comes from the time involved in subroutine calls to the LOG. The author used bit fiddling with Boolean operators to find the exponent, and the execution time decreased to 1.04 seconds.

The evaluation of the Simpson integral presented earlier had problems because of the number of arithmetic operations involved. The statement $X = X + H$ is executed 2^{16} times. Given the size of H, $2.39 \cdot 10^{-5}$, relative to the machine epsilon, $5.96 \cdot 10^{-8}$, there is a considerable accumulation of rounding error in the summation occasioned by the error in X. Sterbenz (1974) discusses way to improve the evaluation of the sum and minimize the error growth. But these improvements are unnecessary if we use one of the more accurate summation methods discussed in this chapter. Both Kahan's trick and cascaded accumulators, for example, calculate the integral as 1.000000, full single-precision accuracy, without further modification to the program.

```
C
C
C    PROGRAM TO DETERMINE THE MACHINE ENVIRONMENT : BASE , PRECISION , ETC.
C
C         PROGRAM MACH—ENV
          IMPLICIT REAL * 8 (A – H , O – Z)
          INTEGER  BETA , T , RND
          DATA RND / 1 / , A / 2.0 / , B / 2.0 / , EPS / 1.0 /
C
C    KEEP DOUBLING A UNTIL ADDING  1 TO IT CAUSES NO CHANGE.
C
          DO  WHILE (( A + 1.0 ) – A . EQ . 1.0 )
              A = 2.0 * A
          END DO
C
C    DETERMINE MACHINE BASE.
C
```

Figure 1.4. Program to reveal floating-point characteristics.

```
      DO  WHILE  (( A + B ) − A .EQ. 0.0 )
           B = 2.0 * B
      END  DO
C
C     FIND OUT IF MACHINE ROUNDS OR CHOPS.
C
      BETA = ( A + B ) − A
      IF (( A + FLOAT ( BETA − 1 )) − A .EQ. 0.0 )  RND = 0
C
C     FIND OUT PRECISION.
C
      T = 1
      A = FLOAT ( BETA )
      DO WHILE (( A +1.0 ) − A . EQ . 1.0 )
           T = T + 1
           A = A * FLOAT ( BETA )
      END  DO
C
C     MACHINE EPSILON IS RECIPROCAL OF A.
C
      EPS = 1.0 / A
C
C     FIND OUT IF GUARD DIGIT IS PRESENT.
C
      BASE = B
      ONEMINUS = ( 0.5 − EPS ) + 0.5
      ULPRADIX = BASE * EPS
      BASEMINUS = BASE − 1.0
      BASEMINUS = ( BASEMINUS − ULPRADIX ) + 1.0
      X = 1.0 − EPS
      Y = 1.0 − ONEMINUS
      Z = 1.0 − X
      XX = BASE − ULPRADIX
      YY = BASE − BASEMINUS
      ZZ = BASE − XX
      IF (( Y . EQ . EPS ) .AND. ( Z . EQ. EPS ) . AND . ( YY . EQ . ULPRADIX ) . AND .
     1 ( ZZ . EQ . ULPRADIX )) THEN
           TYPE * , 'ADD / SUBTRACT GUARD DIGIT PRESENT '
      ELSE
           TYPE * , 'ADD / SUBTRACT GUARD DIGIT ABSENT '
      END IF
C
C     PRINT  RESULTS.
C
      WRITE ( * , 50 ) EPS , BETA , T
50    FORMAT ( '                                            ', /
     1 'MACHINE EPSILON =    ', D22.15 , / ' MACHINE BASE =        ', I4 , /
     2 'MACHINE PRECISION =   ', I4 )
      IF ( RND . EQ . 0 ) THEN
           TYPE *, 'MACHINE CHOPS '
      ELSE
           TYPE *, 'MACHINE ROUNDS '
      END  IF
      END
```

Figure 1.4 (*continued*)

```
                    PROGRAM SUM
                    REAL *8 SD
                    BYTE  D (4), C (4)
                    DIMENSION  X (20000), T (100), ACCUM (64), IX (4)
                    DATA  B/1. 0E38/, NU/8/
                    EQUIVALENCE  (IX, Y), (D (1), A1), (C (1), A2)
                    IZ = 329457
                    SS = 0.0
                    SD = 0.0
                    N = 20000
                    DO  I = 1, N
                       X ( I ) = 100000.0 * (0.5 - RAN ( IZ ))
                    END DO
C  RECURSIVE SUM IN SINGLE PRECISION
                    DO  I = 1, N
                       SS = SS + X ( I )
                    END  DO
                    TYPE *, ' SINGLE PRECISION SUM IS :            ', SS
C  RECURSIVE SUM IN DOUBLE PRECISION
                    DO  I = 1, N
                       SD = SD + DBLE ( X ( I ))
                    END  DO
                    TYPE *, ' DOUBLE PRECISION SUM IS :            ', SS
C  SUMMING WITH KAHAN 'S TRICK
                    S = 0.0
                    S2 = 0.0
                    DO  I = 1, N
                       S2 = S2 + X ( I )
                       R = S + S2
                       S2 = ( S - R ) + S2
                       S = R
                    END  DO
                    TYPE *, ' SUM WITH KAHAN TRICK IS :            ', S
C  BINARY SUMMATION
                    AM = LOG 10 (FLOAT ( N ))/ LOG 10 ( 2.0 )
                    M = AM
                    IF (( AM - M ). GT . 0.01 )  M = M + 1
                    DO  I = 1, M
                      T ( I ) = B
                    END  DO
                    DO  I = 1, N - 1, 2
                      S = S ( I ) + X ( I + 1 )
                      DO  J = 1, M
                        IF ( T ( J ). EQ . B )  GOTO 5
                        S = S + T ( J )
                        T ( J ) = B
                      END  DO
                      T ( J ) = S
                    END  DO
                    S = 0.0
                    DO  J = 1, M
                      IF ( T ( J ). NE . B )  S = S + T ( J )
                    END  DO
                    TYPE *, ' BINARY SUMMATION SUM IS :            ', S
```

Figure 1.5. Program illustrating accurate summation methods.

```
C  SUMMING WITH CASCADED ACCUMULATORS
          DO  I = 1, 64
             ACCUM ( I ) = 0.0
          END  DO
          DO  I = 1, N
            Y = X( I )
            A1 = Y
            D ( 4 ) = '00 'X
            D ( 3 ) = '00 'X
            A2 = X( I ) - A1
            LEV = ( LOG ( ABS ( A1 ) ) / LOG ( 2.0 ) ) + 129
            IEXP = LEV / 4
            ACCUM ( IEXP ) = ACCUM (IEXP) + A1
            LEV = ( ABS ( A2 ) ) / LOG ( 2.0 ) ) + 129
            IEXP = LEV / 4
            ACCUM ( IEXP ) = ACCUM ( IEXP ) + A2
          END  DO
          S = 0.0
          DO  I = 64, 1, -1
            S = S + ACCUM ( I )
          END  DO
          TYPE  * , ' SUM WITH CASCADED ACCUMULATORS IS :  ', S
          END
```

Figure 1.5 (*continued*)

References

Forsythe, G. and Moler, C.B. (1967). *Computer Solution of Linear Algebraic Systems* (Prentice-Hall, Englewood Cliffs, N.J.)

Gentleman, W.M. and Marovich, S.B. (1974). More on Algorithms that Reveal Properties of Floating-Point Arithmetic Units, *Commun. ACM*, **17**, p. 276.

Kahan, W. (1965). Further Remarks on Reducing Truncation Errors, *Commun. ACM*, **8**, p. 40.

Karpinski, R. (1985). Paranoia: A Floating-Point Benchmark, *BYTE*, **10**, No. 2, p. 223.

Kernighan, B.W. and Plauger, P.J. (1974). *The Elements of Programming Style* (McGraw-Hill, New York).

Linnainmaa, S. (1981). Software for Doubled-Precision Floating-Point Computations, *ACM Trans. Math. Software*, **7**, p. 272.

Linz, P. (1970). Accurate Floating-Point Summation, *Commun. ACM*, **13**, p. 361.

Malcolm, M.A. (1971). On Accurate Floating-Point Summation, *Commun. ACM*, **14**, p. 731.

Malcolm, M.A. (1972). Algorithms to Reveal Properties of Floating-Point Arithmetic, *Commun. ACM*, **15**, p. 949.

Scheid, F. (1968). *Theory and Problems of Numerical Analysis* (McGraw-Hill, New York).

Sterbenz, P.H. (1974). *Floating-Point Computation* (Prentice-Hall, Englewood Cliffs, N.J.).

Walker, R.J. (1971). Binary Summation, *Commun. ACM*, **14**, p. 417.

Wolfe, J.M. (1964). Reducing Truncation Errors by Programming, *Commun. ACM*, **7**, p. 355.

CHAPTER 2

Matrices, Norms, and Condition Numbers

2.1. Matrices

Matrix notation greatly facilitates the theoretical discussion of overdetermined systems, although the actual computational steps are often more effectively implemented by other means. Matrices are so familiar that to present a definition of them seems unduly formal, almost a waste of time. Nevertheless, for the sake of completeness we give a definition. A matrix is an array of mn elements, arranged in m rows and n columns,

$$\mathbf{A} = \begin{pmatrix} a_{11} & a_{12} & \cdots & a_{1n} \\ a_{21} & a_{22} & \cdots & a_{2n} \\ \vdots & \vdots & & \vdots \\ a_{m1} & a_{m2} & & a_{mn} \end{pmatrix}. \tag{2.1}$$

The elements may be numbers, real or complex, algebraic expressions, or even other matrices. Matrices obey the multiplication rule

$$\mathbf{C} = \mathbf{A} \cdot \mathbf{B} \quad \Rightarrow \quad c_{ij} = \sum_{k=1}^{n} a_{ik} b_{kj}, \tag{2.2}$$

or multiply the columns of $\mathbf{B}$ by the rows of $\mathbf{A}$. The numbers of elements in a row of $\mathbf{A}$ must equal the number of elements in a column of $\mathbf{B}$ to multiply $\mathbf{A}$ and $\mathbf{B}$.

Matrix multiplication is, in general, not commutative,

$$\mathbf{A} \cdot \mathbf{B} \neq \mathbf{B} \cdot \mathbf{A}, \tag{2.3}$$

but is associative

$$\mathbf{A} \cdot (\mathbf{B} \cdot \mathbf{C}) = (\mathbf{A} \cdot \mathbf{B}) \cdot \mathbf{C}. \tag{2.4}$$

When matrix multiplication does obey the commutative law the matrices involved are called, logically enough, commuting matrices. We shall shortly see examples of some of them.

Certain matrices are so common that they have special names. The null matrix $\mathbf{0}$ contains all zeros. It is a commuting matrix, but little used. Another commuting matrix, the unit matrix

$$I = \begin{pmatrix} 1 & 0 & \cdots & 0 \\ 0 & 1 & & 0 \\ \vdots & \vdots & & \vdots \\ 0 & 0 & & 1 \end{pmatrix}, \tag{2.5}$$

figures prominently in theoretical developments. Both the null and unit matrices, in addition to being commuting, possess two additional properties: they are square, $m = n$, and they are generated; their elements need not be stored in computer memory because they are known and can be generated whenever we need them.

Another commuting square—but not generated matrix—is the inverse. Division of two matrices is not defined, but the inverse matrix satisfies one of the properties of division: if $a = 1/b$, then there exists a number b^{-1} such that $ab^{-1} = 1$; b, of course, must not be zero. With matrices the analogous relation is

$$\mathbf{A} \cdot \mathbf{B} = \mathbf{I} \quad \Rightarrow \quad \mathbf{B} = \mathbf{A}^{-1} \tag{2.6}$$

if $\mathbf{B}$ is nonsingular. A nonsingular matrix is one whose determinant is nonzero. We repeat that the inverse matrix commutes,

$$\mathbf{A} \cdot \mathbf{A}^{-1} = \mathbf{A}^{-1} \cdot \mathbf{A} = \mathbf{I}. \tag{2.7}$$

Even though a matrix must be square to have an inverse, rectangular matrices ($m \neq n$) admit of a kind of inverse, called the pseudoinverse. We shall see what this entity is when we discuss the method of least squares.

The transpose of a matrix $\mathbf{A}$ is formed by interchanging rows and columns: the first row of $\mathbf{A}$ becomes the first column of the transpose $\mathbf{A}^T$, the second row the second column, and so forth. If $\mathbf{A}$ is $m \times n$, $\mathbf{A}^T$ is $n \times m$. Two important types of square matrices are associated with the transpose. If $\mathbf{A}^T = \mathbf{A}$, the square matrix is symmetric: $a_{ij} = a_{ji}$. If $\mathbf{A}^T = \mathbf{A}^{-1}$, the matrix is orthogonal. Orthogonal matrices are intimately involved with the method of least squares.

The elements a_{ii} of a matrix $\mathbf{A}$ are the diagonal elements. If the nondiagonal elements are zero and the diagonal elements nonzero, the matrix is a diagonal matrix. The unit matrix is a special case of the diagonal matrix with each a_{ii} equal to one. If $a_{ij} = 0$ for $|i - j| > m_b$, where m_b is a positive integer, the matrix is a band matrix of bandwidth $2m_b + 1$. (Some authors take m_b as the bandwidth.) If $m_b = 0$ the matrix is diagonal, tridiagonal if $m_b = 1$, pentadiagonal if $m_b = 2$, and so forth. If $a_{ij} = 0$ for $j < i$ the matrix is upper triangular, and lower triangular if $a_{ij} = 0$ for $j > i$.

Although a mathematically rigorous definition of matrix rank is possible, equally illuminating is a more heuristic description. The rank of a matrix is the number of columns or rows that are linearly independent. Consider the 2×2 unit matrix and the 2×2 martrix

$$\begin{pmatrix} 1 & 1 \\ 1 & 1 \end{pmatrix}.$$

The former is of rank two, the latter of rank one. If the rank of a matrix is less than its order, the matrix is singular. That is, if the order of a matrix is $m \times n$ and its rank is less than the smaller of m and n, the matrix is singular.

Unfortunately, the rank of a matrix—even its singularity—is not invariant, but depends on the arithmetic used and the sizes of the elements relative to the arithmetic. If ε is the machine epsilon in single-precision then

$$\begin{pmatrix} 1 & 0.1\varepsilon \\ 1 & 0.2\varepsilon \end{pmatrix}$$

is of rank one, and singular, in single-precision, but of rank two, and non-singular, in double-precision.

Most readers are undoubtedly familiar with the scalar (or dot or inner) product of a $1 \times n$ matrix, called a row vector, with a $n \times 1$ matrix, called a column vector:

$$s = (u_1 \quad u_2 \quad \cdots \quad u_n) \begin{pmatrix} v_1 \\ v_2 \\ \vdots \\ v_n \end{pmatrix} = \sum_{i=1}^{n} u_i v_i. \tag{2.8a}$$

A notationally elegant way to write Eq. (2.8) considers a row vector as the transpose of a column vector and, therefore,

$$s = \mathbf{u}^T \cdot \mathbf{v}. \tag{2.8b}$$

Less familiar is the backwards, or matrix, product of two vectors,

$$\mathbf{u} \cdot \mathbf{v}^T = \begin{pmatrix} u_1 \\ u_2 \\ \vdots \\ u_n \end{pmatrix} (v_1 \quad v_2 \quad \cdots \quad v_n) = \begin{pmatrix} u_1 v_1 & u_1 v_2 & \cdots & u_1 v_n \\ u_2 v_1 & u_2 v_2 & \cdots & u_2 v_n \\ \vdots & \vdots & & \vdots \\ u_n v_1 & u_n v_2 & \cdots & u_n v_n \end{pmatrix}. \tag{2.9}$$

From two vectors we obtain a full $n \times n$ matrix. In fact, Eq. (2.9) is too specific. One vector may be of dimension m and the other of dimension n, giving a backwards product of dimension $m \times n$ or $n \times m$. But, as each row is a multiple of the first, the backwards product is subrank, rank one to be precise.

Before leaving the subject of matrices we should mention, as a historical aside, that Eq. (2.2) represents one way to multiply matrices, but there is another. In the 1930s and 1940s, Cracovians were occasionally used, especially in the astronomical computing literature. Cracovians, introduced by the Polish mathematician and astronomer T. Banachiewicz and named in honor of the Polish city of Cracow, obey the multiplication law

$$\mathbf{C} = \mathbf{A} \cdot \mathbf{B} \quad \Rightarrow \quad c_{ij} = \sum_{k=1}^{n} a_{ik} b_{jk}. \tag{2.10}$$

Instead of multiplying one row of $\mathbf{A}$ into a column of $\mathbf{B}$ we multiply one column of $\mathbf{A}$ into a column of $\mathbf{B}$. For hand calculation Cracovians possess a great advantage over matrices. As anyone who has tried to do it by hand knows, it is almost impossible to multiply two matrices together correctly. The hand multiplication, column by column, of two Cracovians is much less error prone. For systematic work we can make stencils that show only the columns to be multiplied and block out everything else. But for work with a computer, which could not care less if it multiplies row by column or column by column, Cracovians offer no advantage and have fallen into disuse. The author only mentions them so that the reader may recognize a Cracovian if he encounters one in the literature—it is expressed with braces

$$\left\{ \begin{matrix} a_{11} & a_{12} & \cdots & a_{1n} \\ a_{21} & a_{22} & \cdots & a_{2n} \\ \vdots & \vdots & & \vdots \\ a_{m1} & a_{m2} & \cdots & a_{mn} \end{matrix} \right\}, \tag{2.11}$$

and because, as his first scientific paper used Cracovians, he has a nostalgic fondness for them.

2.2. Vector and Matrix Norms

The concept of the norm of a vector is intuitively related to the idea of the length of a vector. The norm is a single number that gives some idea of the size of the elements of the vector which, upon pondering the matter, is just what the length of a vector does. A matrix norm generalizes the idea of length and applies it to an entire matrix of individual lengths of the row or column vectors that constitute the matrix. Both Faddeeva (1959) and Forsythe and Moler (1967) have good discussions of vector and matrix norms.

A vector norm obeys the properties (where the double vertical bars indicate the norm):

$$\|c\mathbf{X}\| = |c| \, \|\mathbf{X}\| \quad \text{for } c \text{ real;} \tag{2.12}$$

if $\mathbf{0}$ is the null vector then

$$\|\mathbf{0}\| = 0 \quad \text{and} \quad \|\mathbf{X}\| > 0; \tag{2.13}$$

$$\|\mathbf{X} + \mathbf{Y}\| \le \|\mathbf{X}\| + \|\mathbf{Y}\|. \tag{2.14}$$

Property (2.14) is called the triangular inequality; it is a generalized Pythagorean theorem. With these properties in mind the norm of an n vector may be defined as

$$\|\mathbf{X}\|_p = (|X_1|^p + |X_2|^p + \cdots + |X_n|^p)^{1/p}. \tag{2.15}$$

If $p = 2$, Eq. (2.15) is nothing more than the Euclidean length of a vector,

$$\|\mathbf{X}\|_2 = \sqrt{X_1^2 + X_2^2 + \cdots + X_n^2}. \tag{2.16}$$

It is common to refer to the norm of a vector as an L_p norm depending on the value of p. The Euclidean length of a vector is the L_2 norm. Other norms widely used are the L_1 (absolute norm),

$$\|X\|_1 = |X_1| + |X_2| + \cdots + |X_n|, \tag{2.17}$$

and the L_∞(max) norm

$$\|X\|_\infty = \max_i |X_i|. \tag{2.18}$$

All of these norms, although distinct, generally agree with one another within an order of magnitude or less. For computational convenience the L_1 and L_∞ norms are often used. They are easier to compute than the L_2 norm and avoid possible difficulties with overflow although they appeal less to our intuitive notion of length. (To avoid the difficulty of overflow we may compute the Euclidean vector norm this way:

$$t = \max_i |X_i|,$$

$$\|X\|_2 = t\sqrt{\left(\frac{X_1}{t}\right)^2 + \left(\frac{X_2}{t}\right)^2 + \cdots + \left(\frac{X_n}{t}\right)^2}.) \tag{2.19}$$

The norm of a matrix is defined so that it obeys properties similar to (2.12)–(2.14). If we consider a matrix $\mathbf{A}$ then, analously to (2.12)–(2.14), we have

$$\|c\mathbf{A}\| = |c|\,\|\mathbf{A}\| \quad (c \text{ real}); \tag{2.20}$$

if $\mathbf{0}$ is the null matrix then

$$\|\mathbf{0}\| = 0 \quad \text{and} \quad \|\mathbf{A}\| > 0; \tag{2.21}$$

$$\|\mathbf{A} + \mathbf{B}\| \le \|\mathbf{A}\| + \|\mathbf{B}\|; \tag{2.22}$$

In addition, with matrices we admit another property

$$\|\mathbf{A} \cdot \mathbf{B}\| \le \|\mathbf{A}\| \cdot \|\mathbf{B}\|. \tag{2.23}$$

The property (2.23) is unnecessary with vectors because the inequality is always obeyed unless the vectors are null.

With the properties (2.20)–(2.23) in mind we may define the norm of a matrix as

$$\|\mathbf{A}\| = \max_{\mathbf{x} \ne \mathbf{0}} \frac{\|\mathbf{A} \cdot \mathbf{x}\|}{\|\mathbf{x}\|}. \tag{2.24}$$

Let us be clear about what Eq. (2.24) is saying. The product $\mathbf{A} \cdot \mathbf{x}$ is a vector whose norm is a number. This number is divided by another nonzero number. We consider the largest ratio of the one number divided by the other as the norm. The properties (2.20) and (2.21) follow directly from the definition (2.24). Equation (2.22) can be demonstrated easily enough by use of (2.24) and property (2.14) of vector norms. Property (2.23) needs a little more work. From (2.24) we infer that in general,

$$\|\mathbf{A}\| \le \frac{\|\mathbf{A} \cdot \mathbf{x}\|}{\|\mathbf{x}\|}. \tag{2.25}$$

Let $\mathbf{x} = \mathbf{B} \cdot \mathbf{y}$ and, from (2.25),

$$\|\mathbf{A}\| \, \|\mathbf{B} \cdot \mathbf{y}\| \leq \|\mathbf{A} \cdot \mathbf{B} \cdot \mathbf{y}\|. \tag{2.26}$$

By use of (2.25) again

$$\|\mathbf{A}\| \, \|\mathbf{B}\| \, \|\mathbf{y}\| \leq \|\mathbf{A} \cdot \mathbf{B} \cdot \mathbf{y}\| \quad \Rightarrow \quad \|\mathbf{A}\| \, \|\mathbf{B}\| \leq \frac{\|\mathbf{A} \cdot \mathbf{B} \cdot \mathbf{y}\|}{\|\mathbf{y}\|}. \tag{2.27}$$

Finally, we infer, considering once again (2.25)

$$\|\mathbf{A}\| \, \|\mathbf{B}\| \leq \|\mathbf{A} \cdot \mathbf{B}\|,$$

property (2.23).

If in Eq. (2.25) we take $\|\mathbf{x}\| = 1$ then we may derive three useful matrix norms. We shall merely state them here. For the derivations, see Faddeeva (1959, pp. 56–60).

The L_2 norm of a matrix $\mathbf{A}$ is

$$\|\mathbf{A}\|_2 = \sqrt{\lambda_{max}}, \tag{2.28}$$

where λ_{max} is the largest eigenvalue of $\mathbf{A}^T \cdot \mathbf{A}$. Presumably, most readers are familiar with eigenvalues (sometimes called proper values) and eigenvectors. But for the sake of completeness we shall present some definitions. An eigenvalue λ is a number with the property

$$\mathbf{A} \cdot \mathbf{x} = \lambda \mathbf{x}. \tag{2.29}$$

(If all λ are real and positive the matrix is positive definite.) An eigenvector $\mathbf{x}$ is any nonzero solution vector of the system

$$(\mathbf{A} - \lambda \mathbf{I})\mathbf{x} = 0. \tag{2.30}$$

By analogy with the L_2 vector norm we might have expected that the L_2 matrix norm of an $m \times n$ matrix $\mathbf{A}$ would be

$$\|\mathbf{A}\| = \left(\sum_{i=1}^{m} \sum_{j=1}^{n} a_{ij}^2 \right)^{1/2}. \tag{2.31}$$

Such a norm, in fact, exists and is called the Schur or Euclidean norm. For vector norms we may refer indifferently to the L_2 or Euclidean norm as they are the same, but with matrices we must be more careful with our terminology. The Euclidean matrix norm is easier to calculate than the L_2 matrix norm, but the latter possesses more fundamental properties, as we shall see when discussing the method of least squares.

The L_1 and L_∞ matrix norms are closely related. For the former

$$\|\mathbf{A}\|_1 = \max_{j} \sum_{i=1}^{n} |a_{ij}|, \tag{2.32}$$

and for the latter

$$\|\mathbf{A}\|_\infty = \max_{i} \sum_{j=1}^{n} |a_{ij}|. \tag{2.33}$$

In words, the L_1 matrix norm is the maximum of the sums of the absolute values of the elements of the columns of $\mathbf{A}$. For the L_∞ norm substitute row for column. From (2.32) and (2.33) we infer

$$\|\mathbf{A}\|_1 = \|\mathbf{A}^T\|_\infty. \tag{2.34}$$

As with vector norms the matrix norms differ little among themselves. A simple example:

$$\mathbf{A} = \begin{pmatrix} 4.1 & 9.7 \\ 2.8 & 6.6 \end{pmatrix}.$$

By inspection $L_1 = \|\mathbf{A}\|_1 = 16.3$ and $L_\infty = 13.8$. The Schur norm requires a little computing, for which a hand calculator suffices to give $L_E = 12.7$ (the subscript E stands for Euclidean). To calculate the L_2 norm again a hand calculator will do. From

$$\mathbf{A}^T \cdot \mathbf{A} = \begin{pmatrix} 24.65 & 58.25 \\ 58.25 & 137.65 \end{pmatrix}$$

we obtain the secular equation $\lambda^2 - 162.3\lambda + 0.01 = 0$ and, therefore, $\sqrt{\lambda_{max}} = 12.7$. (The author hastens to add that calculating eigenvalues directly from the secular equation that arises from the expansion of Eq. (2.30) is the worst way to do it. But for this simple example the secular equation suffices.)

2.3. The Condition Number

If the matrix we are working with is singular our linear system is generally of little use. Given the discussion of Chapter 1 we should realize that closeness to singularity, in the compromised world of floating-point arithmetic, will also plague us. But how do we measure "closeness to singularity"? We sometimes hear that a small determinant of the matrix indicates near singularity. This assertion, however, is incorrect. Certainly, if the determinant is exactly zero, the matrix is singular. But with floating-point arithmetic we rarely encounter an exact zero, just a small number that may or may not approximate zero. A number of the order of the machine epsilon, for example, may result from calculations for the determinant of a singular matrix. Or it may be a small number in its own right—consider the size of Planck's constant in c.g.s. units: of the order of 10^{-27}—having nothing to do with the singularity of a matrix.

Let us clarify the preceding somewhat. Suppose that our machine's single-precision epsilon is of the order of 10^{-8}. The determinant of the matrix

$$\begin{pmatrix} 2 \cdot 10^{-4} & 0 \\ 0 & 2 \cdot 10^{-4} \end{pmatrix}$$

is $4 \cdot 10^{-8}$, but the matrix is not even close to singularity; its columns are independent. The matrix, in fact, is nothing more than a scaled unit matrix.

On the other hand, the matrix

$$\begin{pmatrix} 1.00000001 & 0.99999999 \\ 0.99999999 & 1.00000001 \end{pmatrix},$$

whose determinant is also $4 \cdot 10^{-8}$, is nearly singular; the two columns nearly coincide. The size of the determinant, $4 \cdot 10^{-8}$, gives no indication of the near singularity of one matrix and the perfect behavior of another.

To find out how we can measure closeness to singularity let us start with the linear system

$$A \cdot X = b. \tag{2.35}$$

If Eq. (2.35) is a pure mathematical expression it has an exact solution X, for nonsingular A, and nothing further need be said. But if the system arises from an experiment the situation is more interesting. A will contain coefficients derived from a mathematical model. The coefficients may be safely considered error free. Even if the model is incorrect, once it is adopted the coefficients are known exactly. b, on the other hand, comes from the data and is corrupted by experimental errors. Given the errors in b, what effect have they on the solution X?

Let b consist of an (unknown) exact part, b_0, plus a perturbation Δb caused by the errors. X likewise consists of an exact X_0 plus an error term ΔX that reflects the experimental errors in b. From Eq. (2.35) we have

$$A \cdot (X_0 + \Delta X) = b_0 + \Delta b. \tag{2.36}$$

By cancelling out the exact part $A \cdot X_0 = b_0$ we obtain

$$A \cdot \Delta X = \Delta b \quad \Rightarrow \quad X = A^{-1} \cdot \Delta b. \tag{2.37}$$

From property (2.23) of the matrix norm we have

$$\|\Delta X\| \leq \|A^{-1}\| \cdot \|\Delta b\| \tag{2.38}$$

and

$$\|b_0\| \leq \|A\| \cdot \|X_0\|. \tag{2.39}$$

Equation (2.38) combines with Eq. (2.39) to give

$$\frac{\|\Delta X\|}{\|X_0\|} \leq \|A\| \cdot \|A^{-1}\| \cdot \frac{\|\Delta b\|}{\|b_0\|}. \tag{2.40}$$

The term $\|A\| \cdot \|A^{-1}\|$ in Eq. (2.40) is interesting. Suppose we use the L_2 norm. The norm of A is $\sqrt{\lambda_{max}}$ and the norm of A^{-1} is $\sqrt{\lambda_{min}^{-1}}$. Their product, $\lambda_{max}^{1/2}/\lambda_{min}^{1/2}$, reveals that the term is a error magnification factor. At best, λ_{max} equals λ_{min} and the term, being unity, neither magnifies nor diminishes the relative error of the norm of b. But generally, $\lambda_{max} > \lambda_{min}$ and the term amplifies the error of the norm of b. In the worst case, a singular matrix, $\lambda_{min} = 0$ and the term blow up. The relative error of X's norm is infinite. In other words, the solution is unknown. It is common to refer to the term

$\|\mathbf{A}\| \cdot \|\mathbf{A}^{-1}\|$ as the condition number of the matrix $\mathbf{A}$, COND(A):

$$\text{COND}(\mathbf{A}) = \|\mathbf{A}\| \cdot \|\mathbf{A}^{-1}\|, \tag{2.41}$$

with the property $1 \leq \text{COND}(\mathbf{A}) \leq \infty$. Well and fine, but what about the other matrix norms? In Eq. (2.24) we can substitute min instead of max, always keeping in mind that $\|\mathbf{X}\| \neq 0$. At best max equals min, but is usually larger. Regardless of the norm used the minmum for COND(A) is one. Likewise, COND(A) can go to infinity because the minimum of the norm may be zero even if $\|\mathbf{X}\| \neq 0$: suppose that $\|\mathbf{X}\| = 1$ and $\mathbf{A}$ is the null matrix.

To help clarify some of these concepts take as an example the linear system,

$$\begin{pmatrix} 5.5 & 8.0 & 3.7 \\ 10.2 & 15.0 & 7.0 \\ 16.9 & 25.5 & 11.5 \end{pmatrix} \begin{pmatrix} X_1 \\ X_2 \\ X_3 \end{pmatrix} = \begin{pmatrix} 17.2 \\ 32.2 \\ 53.9 \end{pmatrix}, \tag{2.42}$$

whose solution is $X_1 = X_2 = X_3 = 1$. To one decimal the inverse of the matrix is

$$\mathbf{A}^{-1} = \begin{pmatrix} 10.3 & -4.1 & -0.9 \\ -1.7 & -1.2 & 1.3 \\ -11.4 & 8.7 & -1.6 \end{pmatrix}. \tag{2.43}$$

To simplify the calculations we shall use the L_1 norm. The L_1 norm of the matrix of Eq. (2.42) is 48.5 and of its inverse, (2.43), 23.4, for a condition number of 1,134.9. A small variation in $\mathbf{b}$ should produce a large variation in $\mathbf{X}$. Add 0.1 to the components of $\mathbf{b}$. We have for the relative error in the L_1 norm of $\mathbf{b}$

$$\frac{\|\Delta\mathbf{b}\|_1}{\|\mathbf{b}\|_1} = \frac{0.3}{103.3},$$

about 0.3%. The new solution vector is $X_1 = 1.54$, $X_2 = 0.83$, $X_3 = 0.58$. The relative error in the norm of the solution is

$$\frac{\|\Delta\mathbf{X}\|_1}{\|\mathbf{X}\|_1} = \frac{1.13}{3},$$

a relative error of 38%. As expected the determinant of the matrix, -0.58, tells us nothing; it is useless.

But suppose that our system involves no experimental data. What if we merely want to invert a matrix? The condition number is still relevant. The inversion of an $n \times n$ matrix is nothing more than the solution of n linear systems. The columns of the inverse $\mathbf{A}^{-1}$ of a matrix $\mathbf{A}$ are solutions to the equations

$$\mathbf{A} \cdot \mathbf{X}_i = \mathbf{e}_i, \tag{2.44}$$

where $\mathbf{e}_i$ are unit vectors such that $\mathbf{e}_1^T = (100\ldots)$, $\mathbf{e}_2^T = (010\ldots)$, ..., $\mathbf{e}_n^T = (00\ldots1)$. In Chapter 13 of their book, Forsythe and Moler (1967) prove that solving a linear system by Gaussian elimination can involve the loss of COND(A) digits in the solution, caused by denormalizations and consequent loss of significant bits in the floating-point operations.

Although Forsythe and Moler's argument is involved, a simple argument shows how the condition number is important for matrix inversion. Multiplication of the original matrix by its inverse should give a unit matrix. Because of the precision loss caused by floating-point operation the product differs from the unit matrix. We define a matrix of residuals

$$\mathbf{R} = \mathbf{I} - \mathbf{A} \cdot \mathbf{A}^{-1}. \tag{2.45}$$

By the properties of the matrix norm we have

$$\|\mathbf{R}\| = \|\mathbf{I} - \mathbf{A} \cdot \mathbf{A}^{-1}\| \simeq \left| \|\mathbf{I}\| - \|\mathbf{A}\| \cdot \|\mathbf{A}^{-1}\| \right|$$

or

$$\|\mathbf{R}\| \simeq |1 - \text{COND}(\mathbf{A})| \tag{2.46}$$

because the norm of a unit matrix is always unity. If the condition number is unity the norm of the residual matrix is zero; the residuals are, as they should be, zero. But as COND(A) increases the residuals become larger. If COND(A) becomes too large the matrix cannot be inverted. From Forsythe and Moler's rigorous argument this happens when COND(A) is of the order of the reciprocal of the machine epsilon.

To see the effect of the condition number on matrix inversion let us do a simple example, so simple that a Sharp EL-5500 II hand calculator/computer, with machine epsilon of $5 \cdot 10^{-10}$, suffices. Our matrix is

$$\mathbf{A} = \begin{pmatrix} 1 & 1 + \sqrt{\varepsilon} \\ 1 & 1 + 2\sqrt{\varepsilon} \end{pmatrix} = \begin{pmatrix} 1 & 1.000022361 \\ 1 & 1.000044721 \end{pmatrix}$$

with inverse

$$\mathbf{A}^{-1} = \begin{pmatrix} 44724.71919 & -44723.71919 \\ -44722.71914 & 44722.71914 \end{pmatrix}.$$

We calculate the condition number as $9 \cdot 10^4$ in the L_1 norm and expect to lose up to five digits in the product of A by $\mathbf{A}^{-1}$. In fact

$$\mathbf{A} \cdot \mathbf{A}^{-1} = \begin{pmatrix} 1.000005300 & 0 \\ 0 & 0.999994700 \end{pmatrix};$$

two of the elements are perfect and we have lost five digits in the other two.

With the condition number we can give a precise meaning to the loose term "ill-conditioned". A matrix is ill-conditioned if its condition number is high with respect to the precision of the floating-point arithmetic being used. On a computer like a VAX or IBM 370, a matrix whose condition number is 10^6 would be ill-conditioned with respect to single-precision arithmetic: we could lose all but one or two of the significant decimal digits of a number. A matrix with COND(A) = 10^9 would be uninvertiable in single-precision, but only moderately ill-conditioned in double-precision. Should the condition number be 10^{50} then no precision, single, double, or extended, can do anything with the matrix.

The definition of condition number used so far, Eq. (2.41), suffers a defect. A matrix such as

$$\begin{pmatrix} 10^{30} & 0 \\ 0 & 10^{-30} \end{pmatrix},$$

a scaled unit matrix, seems well-conditioned, but has COND(A) = 10^{60}. For this reason Skeel (1979) proposed an alternative definition,

$$\text{COND(A)} = \| \, |A| \cdot |A^{-1}| \, \|, \tag{2.47}$$

where the notation means that we should multiply the absolute values of the elements of A and its inverse. With the definition (2.47) the condition number of a diagonal matrix is always unity.

We should distinguish between the condition number of a matrix and that of a linear system. The matrix

$$\begin{pmatrix} 1 & 2 & 0 \\ 3 & 4 & 0 \\ 0 & 0 & 0 \end{pmatrix} \tag{2.48}$$

is singular, but the linear system

$$\begin{pmatrix} 1 & 2 & 0 \\ 3 & 4 & 0 \\ 0 & 0 & 0 \end{pmatrix} \begin{pmatrix} X_1 \\ X_2 \\ 0 \end{pmatrix} = \begin{pmatrix} 1 \\ 1 \\ 0 \end{pmatrix} \tag{2.49}$$

has the solution $X_1 = -1$, $X_2 = 1$. The last equation merely adds the trivial identity $0 \cdot 0 = 0$. Skeel (1978) introduces, as the condition number of a linear system,

$$\text{COND(A, b)} = \| \, |A| |A^{-1}| \cdot |X| + |A^{-1}| \cdot |b| \, \|. \tag{2.50}$$

The notation makes it clear that the condition number of a linear system depends on the right-hand side **b**.

To see why Eq. (2.50) is a good definition let us calculate the condition number of Eq. (2.49). A difficulty arises immediately with the inverse matrix A^{-1}. Because the matrix (2.48) is singular its inverse cannot be calculated. To surmount this difficulty we resort to a trick common in mathematics when singularities present themselves: introduce a small quantity δ in place of zero and at the opportune moment let it go to zero. In place of (2.48) we use

$$A = \begin{pmatrix} 1 & 2 & \delta \\ 3 & 4 & \delta \\ \delta & \delta & \delta \end{pmatrix}, \tag{2.51}$$

which has the inverse

$$A^{-1} = \begin{pmatrix} -\tfrac{1}{2}(4 - \delta) & 1 - \tfrac{1}{2}\delta & 1 \\ \tfrac{1}{2}(3 - \delta) & -\tfrac{1}{2}(1 - \delta) & -1 \\ \tfrac{1}{2} & -\tfrac{1}{2} & 1/\delta \end{pmatrix}. \tag{2.52}$$

Because of the $1/\delta$ term, A^{-1} blows up as $\delta \to 0$. But upon multiplying $|A|$

by $|A^{-1}|$ we find that the obnoxious $1/\delta$ term is multiplied by δ, rendering it harmless. When $|A^{-1}|$ is multiplied by $|b|$ the term is zeroed by the zero in **b**. If the zero in **b** were not present the condition number of Eq. (2.50) would also blow up, as it should because the last equation of the linear system would be an inconsistent $0 \cdot 0 =$ finite quantity. Upon letting $\delta \to 0$, we find that the condition number of Eq. (2.49) in the L_1 norm is 30.

This example also shows that, unfortunately, the definition (2.47) suffers a defect. Upon introducing the finite δ in matrix (2.48) and letting it go to zero, we nevertheless obtain a finite condition number for the matrix. Although (2.47) alleviates the problem of an unrealistic condition number for diagonal matrices, it fails for singular matrices if we introduce the singular matrix by the trick used in the discussion of definition (2.50). To prevent this, we should not introduce the trick of multiplying out the singularity into definition (2.47).

The definitions of the condition number seem to imply that we must explicitly calculate the inverse matrix, a distinct disadvantage if our only interest is solving a linear system. The solution of a linear system by matrix inversion involves much more work than a direct solution of the linear system. Suppose that we wish to solve the triangular system

$$\begin{pmatrix} 1 & 2 \\ 0 & 3 \end{pmatrix}\begin{pmatrix} X_1 \\ X_2 \end{pmatrix} = \begin{pmatrix} 5 \\ 6 \end{pmatrix}. \tag{2.53}$$

X_2 follows by division of six by two, one arithmetic operation. X_1 is obtained with the calculated X_2, multiplied by two, subtracted from five, and divided by one; a further four arithmetic operations and an exact solution of $X_1 = 1$ and $X_2 = 2$.

Had we opted for matrix inversion we would find that, upon applying Eq. (2.44), we need eight multiplications, four subtractions, and four divisions to arrive at (using a Sharp EL-5500 II hand calculator/computer)

$$X = \begin{pmatrix} 1 & -0.666666666 \\ 0 & 0.333333333 \end{pmatrix}\begin{pmatrix} 5 \\ 6 \end{pmatrix}. \tag{2.54}$$

An additional four multiplications and two additions result in $X_1 = 1.000000000$, $X_2 = 1.999999998$. Altogether more arithmetic labor and less precision in the solution.

Another objection to using matrix inversion if out only interest is solving a linear system arises from consideration of the sparcity of the linear system. With practical problems the size of the system may be large, but with many coefficients equal to zero. Such a system is called sparse and is discussed in the next chapter. Unfortunately, the inverse of a sparse matrix may not, and generally is not, sparse. As an example the matrix

$$\begin{pmatrix} 4 & 1 & 2 & 0.5 & 2 \\ 1 & 0.5 & 0 & 0 & 0 \\ 2 & 0 & 3 & 0 & 0 \\ 0.5 & 0 & 0 & 0.625 & 0 \\ 2 & 0 & 0 & 0 & 16 \end{pmatrix}$$

is nearly 50% sparse; out of twenty-five elements, twelve are zero. But its inverse is

$$
\begin{pmatrix}
60 & -120 & -40 & -48 & -7.5 \\
-120 & 242 & 80 & 96 & 15 \\
-40 & 80 & 27 & 32 & 5 \\
-48 & 96 & 32 & 40 & 6 \\
-7.5 & 15 & 5 & 6 & 1
\end{pmatrix}.
$$

The symmetry of the original matrix is preserved, but all of its zero elements have suffered fill-in. Should we wish to take advantage of sparse matrix techniques for the original matrix, these techniques could not be used for the inverse.

The moral of the preceding two paragraphs is clear: unless we have an explicit need for the inverse matrix we should dispense with it.

But, if we do not have an inverse matrix how can we obtain a condition number? With definitions (2.47) and (2.50) there is no way out of the conundrum: we need an inverse matrix. But with definition (2.41) there is an easy way that involves no matrix inverse. We can estimate the condition number from the square root of the ratio of the largest to the smallest eigenvalue of $A^T \cdot A$. Nor is it necessary to calculate *all* of the eigenvalues. The power method, also known as the Rayleigh method, comes to our aid. This is an iterative procedure that converges to the largest eigenvalue of $A^T \cdot A$. Having the largest eigenvalue we subtract it from the main diagonal of $A^T \cdot A$ and the procedure converges to the smallest eigenvalue. We take the square root of their ratio as the condition number.

As a concrete example suppose that we wish to estimate the condition number of the matrix in Eq. (2.42). We start by forming

$$
A^T \cdot A = \begin{pmatrix}
419.90 & 627.95 & 286.10 \\
627.95 & 939.25 & 427.85 \\
286.10 & 427.85 & 194.94
\end{pmatrix}. \tag{2.55}
$$

just about any nonzero starting vector will do. For automatic computation, a vector of ones, subsequently normalized to unity, is as good as any. With matrix (2.55) we start the iterations with the vector $X^T = (1/\sqrt{3}, 1/\sqrt{3}, 1/\sqrt{3})$. Because the condition number is not needed to great accuracy, single-precision is sufficient. After each multiplication of the current X by matrix (2.55) the new X should be renormalized to unity. The normalization constant is the current approximation to the eigenvalue. Although not needed the X's converge to the eigenvector associated with the eigenvalue.

On a VAX-11/780 the iterations converge to full single-precision accuracy, actually overkill as three decimals would suffice, after five passes to

$$
1554.017 \begin{pmatrix}
0.5197941 \\
0.7774212 \\
0.3541616
\end{pmatrix}.
$$

To find the minimum eigenvalue subtract 1554 from the main diagonal of (2.55), a process technically called shifting. Shifting isolates the smallest eigenvalue so that the iterates converge to it. After four iterations we find

$$5.6884766 . 10^{-2} \begin{pmatrix} 0.2710486 \\ -0.5432277 \\ 0.7946297 \end{pmatrix}.$$

The condition number of the matrix of Eq. (2.42) in the L_2 norm is $(1544.017/5.6884766 \cdot 10^{-2})^{1/2} = 176$, obtained without the need to invert a matrix. But the condition number of 176 seems small compared with the 1134.9 calculated previously from the L_1 norm. We have stated that the condition numbers from the various norms should be fairly close. Are we giving a broad interpretation to the term "fairly close"? Not really. The smallest eigenvalue of matrix (2.55) is really $3.08 \cdot 10^{-3}$, giving a condition number of 710. The three eigenvalues of the matrix are 1554, 0.07, and 0.003. The largest is well separated from the other two, and the power method rapidly converges to a good value for the largest eigenvalue. The other two, however, cluster together, derailing the power method and causing it to converge to a value between 0.07 and 0.003.

Rather than becoming alarmed and opting for an accurate and computationally expensive algorithm that finds *all* of the eigenvalues, or looking for an example where the power method neatly converges to the smallest eigenvalue—selecting another example when a given example has difficulties strikes the author as, if not dishonest, at least intellectually sleazy—we should reflect a little. We do not need the condition number to great accuracy anyway, probably something within an order of magnitude of a condition number calculated with great accuracy is sufficient. Whether our condition number is 176, 710, or 1,135 we know that about three or four digits will be last in solving a linear system or inverting a matrix. This may, or may not, be sufficient precision for the problem at hand. If not we should switch to double-precision.

In summary, the power method affords a way to compute the condition number of a matrix that does not require an explicit inverse matrix. This condition number tells us how much precision we are likely to lose in solving a linear system or, if the number is high, whether the system can even be solved with the precision of arithmetic being used. This information comes from the condition number, the determinant of the system being a relatively useless quantity.

References

Faddeeva, V.N. (1959). *Computational Methods of Linear Algebra* (Dover, New York).
Forsythe, G. and Moler, C.B. (1967). *Computer Solution of Linear Algebraic Systems* (Prentice-Hall, Englewood Cliffs, N.J.).
Skeel, R.D. (1979). Scaling for Numerical Stability in Gaussian Elimination, *J. ACM*, **26**, p. 494.

CHAPTER 3

Sparse Matrices

3.1. Introduction

A matrix is sparse if a high percentage of its elements are zero. "High", of course, is a relative term, but if about 50% or more of the elements are zero the matrix may be considered sparse. With this percentage of null elements, use of data structures other than the standard two-dimensional array may result in a substantial saving of space or execution time, space because the storage of the zero elements is suppressed and time because we do no operations with them. Consider scaling the columns of an $n \times n$ matrix $\mathbf{A}$. Mathematically, we do this by postmultiplying $\mathbf{A}$ by a diagonal matrix $\mathbf{S}$ of scale factors,

$$\mathbf{A}' = \mathbf{A} \cdot \mathbf{S}. \tag{3.1}$$

But no sane programmer would store $\mathbf{S}$ as an $n \times n$ matrix. He would store it as a vector, thus using only n memory locations. Likewise, he would not use a matrix multiplication routine for the product $\mathbf{A} \cdot \mathbf{S}$. He would multiply all of the elements in the first column of $\mathbf{A}$ by the first scale factor, the second column by the second scale factor, and so forth. In this way only n^2 arithmetic operations, rather than n^3 as with matrix multiplication, would be needed.

Sparse matrix techniques are many and varied. Entire books are devoted to them. George and Liu (1981), for example, have written a book on just the solution of sparse positive definite systems. We cannot go into such detail, but hope to give the reader at least an overview of some typical sparse matrix techniques and present some simple routines to clarify the discussion. The reader will also learn that sparse matrix techniques are a mixture of art and science, with the former predominating. As an inducement to the study of these techniques, which become rather involved, the author will give a concrete example of the savings he once experienced when using a sparse matrix technique. He was involved in some research that used a solution in the L_1 norm of an overdetermined system (the L_1 algorithm will be presented in a later chapter). The size of the matrix, 21,365 × 41 in double-precision, was considerably larger than the 1 MB of physical memory available on the

VAX-11/780 used for the calculations. A total of $12\frac{1}{2}$ hours of CPU time was needed to obtain the L_1 solution. This seemed excessive, especially as the author wanted to obtain solutions for a number of right-hand sides. But the matrix was only 41% dense, and upon recasting the algorithm to take advantage of the sparseness, which required a fair amount of effort, the CPU time decreased to 6 hours, a factor of two savings in execution time.

Sparse matrix techniques fall into two broad categories: the null elements follow a distinct pattern, as with a triangular matrix, for example; the null elements occur in random locations. The first is by far the easier to deal with. The array is the best data structure to use because fill-in, should it occur, is predictable: the product of two upper triangular matrices is upper triangular; the product of a low triangular matrix with an upper triangular matrix is a full matrix; the backwards product of two vectors is a matrix. For the second category, the array, although still useful, loses much of its attractiveness because, unless the amount of fill-in can be accurately assessed in operations such as matrix multiplication and inversion, the array may be grossly over-dimensioned or may overflow. To obviate the need for judging fill-in, dynamic data structures, such as the linked list, may be employed. Unfortunately, not all programming languages, and FORTRAN and BASIC are in this group, incorporate dynamic structures. In this chapter examples will be based on FORTRAN and, when dynamic structures are needed, on C. BASIC, a dubious language at best for numerical work, is particularly inept for sparse matrices—its poor memotechnic capability, among other things, is a real handicap—and will be avoided.

3.2. Sparse Techniques for Null Elements Following a Pattern

A trivial example of a sparse system where the zero elements follow a distinct pattern is a diagonal matrix. A vector, or one-dimensional array, would be used to hold the diagonal elements. All matrix operations, such as forming the transpose or finding the inverse, would need no further storage locations, just the vector locations already assigned.

As a nontrivial, but still not too complicated, example, consider an upper triangular matrix

$$\mathbf{A} = \begin{pmatrix} a_{11} & a_{12} & \cdots & a_{1n} \\ 0 & a_{22} & \cdots & a_{2n} \\ \vdots & & & \vdots \\ 0 & 0 & \cdots & a_{nn} \end{pmatrix}. \tag{3.2}$$

Storing $\mathbf{A}$ by the use of a two-dimensional array wastes $n(n-1)/2$ locations on the zeros. And if $\mathbf{A}$ were submitted to a general matrix inversion routine, time would be wasted doing unnecessary calculations with the zero elements. Because the zeros follow a predictable pattern we may profitably employ a vector to contain the nonzero elements. These may be stored either row-wise or column-wise. With the latter arrangement we represent **A** as the vector

$$\mathbf{A} = \begin{pmatrix} a_{11} \\ a_{12} \\ a_{22} \\ \vdots \\ a_{1n} \\ \vdots \\ a_{nn} \end{pmatrix}. \tag{3.3}$$

Our first problem is how to locate a given element a_{ij}. Because computer memory is linear, even two-dimensional arrays are stored internally in a linear arrangement by a mapping of the two indices i and j to one index k. If the two-dimensional array is to be stored by column, as FORTRAN prefers to do it, and the dimensions of the array are m rows and n columns, the mapping function is $k = m(j - 1) + i$. Languages that prefer to store arrays row-wise, such as C and PL/I, would use $k = n(i - 1) + j$.

Because the elements in Eq. (3.3) are stored column-wise we shall employ the mapping function $k = n(j - 1) + i$, which must be modified to allow for the first column's $n - 1$ elements to be missing, the second's $n - 2$ elements, and so forth. This can be done by use of the mapping function

$$k = n(j - 1) + i - \sum_{l=1}^{j-1} (n - l). \tag{3.4a}$$

High-level language like FORTRAN and C easily implement an equation like (3.4a) via a function call, but it can be simplified further, making a call unnecessary.

$$k = nj - n + i - (j - 1)n + \sum_{l=1}^{j-1} l,$$

$$k = nj - n + i - nj + n + \frac{(j - 1)(j - 1 + 1)}{2}, \tag{3.4b}$$

$$k = \frac{j(j - 1)}{2} + i,$$

which is independent of n.

With the compressed representation of the matrix (3.3) and the mapping function (3.4) many common matrix operations, such as the matrix product, are easy. Suppose we need to transpose an upper triangular matrix. Were we to use a separate array for the transpose we would scan for the elements in each row of **A** and store them consecutively in the array for $\mathbf{A}^{\mathrm{T}}$. The coding would look something like

```
DO I = 1, N
  DO J = I, N
    ATRANS(I * (I − 1)/2 + J) = A(J * (J − 1)/2 + I).
```

Although in practice we sometimes want to store the transpose in a separate array, it is more common to use the same array for both matrix and transpose. With the compressed representation of Eq. (3.3) our first response might be to try something like

$$\text{DO } I = 1, N$$
$$\text{DO } J = I, N$$
$$A(I * (I - 1)/2 + J) = A(J * (J - 1)/2 + I),$$

but this fails. To understand why, consider a 4×4 upper triangular matrix and look at where the transposed elements go:

Original	Transpose
a_{11}	a_{11}
a_{12}	a_{12}
a_{22}	a_{13}
a_{13}	a_{14}
a_{23}	a_{22}
a_{33}	a_{23}
a_{14}	a_{24}
a_{24}	a_{33}
a_{34}	a_{34}
a_{44}	a_{44}

As we scan the first row of A elements a_{11} and a_{12} retain their locations, but a_{13} goes to the third location, wiping out a_{22}. Therefore, a_{22} after transposing will really be a_{13}. Other elements will also unwittingly be modified. We must look for some other mechanism.

One possible scheme would be to use an auxiliary array of size n—4 in this instance—to contain the elements of the longest row: a_{11}, a_{12}, a_{13}, a_{14}. Replace these elements by zeros as they are picked off the array. At the end of the row scan shift the remaining elements, starting with a_{22}, downward by an amount large enough to create n zeros at the beginning of the array where the row elements in the temporary array will be placed. Something similar would be done for subsequent rows. But this scheme will involve data movements, with deleterious effects on execution time for large n. The indexing to locate a shifted element is also likely to be complicated.

Fortunately, a simple change of attitude on our part resolves the problem completely. If we make judicious use of our mapping function (3.4) no data movements at all are required. The transpose of an upper triangular matrix is lower triangular. In a program its elements would be accessed by coding like

$$\text{DO } J = 1, N$$
$$\text{DO } I = J, N$$
$$B = A(I, J).$$

The same effect is achieved with the compact representation of **A** by

$$
\begin{aligned}
&\text{DO } J = 1, N \\
&\quad\text{DO } I = J, N \\
&\qquad B = A(I * (I - 1)/2 + J).
\end{aligned}
$$

We simple reverse the i and j indices in our mapping function. What could be easier?

Let us now think of something a little more complicated, namely the inversion of a triangular matrix. One factor works in our favor immediately: the inverse of an upper triangular matrix is upper triangular; there is no fill-in. (If this assertion appears dubious to the reader he should work out the mathematics by considering the inverse to be a full matrix. From the relation

$$
\begin{pmatrix}
a_{11} & a_{12} & \cdots & a_{1n} \\
0 & a_{22} & \cdots & a_{2n} \\
\vdots & & & \\
0 & 0 & \cdots & a_{nn}
\end{pmatrix}
\begin{pmatrix}
a_{11}^{-1} & a_{12}^{-1} & \cdots & a_{1n}^{-1} \\
a_{21}^{-1} & a_{22}^{-1} & \cdots & a_{2n}^{-1} \\
\vdots & & & \\
a_{n1}^{-1} & a_{n2}^{-1} & \cdots & a_{nn}^{-1}
\end{pmatrix}
=
\begin{pmatrix}
1 & 0 & \cdots & 0 \\
0 & 1 & & 0 \\
\vdots & \vdots & & \vdots \\
0 & 0 & & 1
\end{pmatrix}
$$

it will become evident that the subdiagonal elements of the inverse must be zero.)

By remembering that **A** is upper triangular and using the identity $\mathbf{A} \cdot \mathbf{A}^{-1} = \mathbf{I}$, we can derive, without too much difficulty, formulas to calculate the inverse elements:

$$
a_{ii}^{-1} = \frac{1}{a_{ii}}, \qquad i = 1, \dots, n
$$

$$
a_{ij}^{-1} = -a_{jj}^{-1} \sum_{k=1}^{j-1} a_{ik}^{-1} a_{kj}, \qquad j = i+1, \dots, n, \quad i = 1, \dots, n-1. \tag{3.5}
$$

Equations (3.5) possess the advantage that as the inverse elements are calculated they overwrite elements of the original matrix that are no longer needed for the ensuing calculations. **A** is progressively destroyed and $\mathbf{A}^{-1}$ occupies the space previously reserved for **A**.

Figure 3.1 shows a short FORTRAN program, which should be largely self-explanatory, for the in-place inversion of an upper triangular matrix.

The ideas sketched here for the upper triangular matrix may be applied in analogous fashion to other sparse systems where the null elements follow a definite pattern. We consider a suitable storage arrangement for the nonzero elements. A band matrix, for example, could be stored in a one-dimensional array as a sequence of diagonals; other arrangements are also possible. Then we find a suitable mapping function to locate an element a_{ij} in the compressed representation. The complexity of the matrix operations will depend on the particular system. Fortunately, fill-in will either not occur—the inverse of a unit matrix is a unit matrix—or will be predictable—the backwards product of two vectors is a full matrix, although of rank one, so that the array, even the static array of FORTRAN, is the ideal data structure to use.

3.3. Sparse Techniques with Null Elements in Random Locations

When the null elements appear in random locations our first problem is to find a suitable scheme for storing and identifying the nonzero elements. Because of the randomness of the zero elements we can no longer use a mapping function to transform from i and j indices to an index that uniquely identifies the element in a one-dimensional array. There are various possibilities for locating a nonzero element, all of which rely on subsidiary arrays or, with dynamic data structures, on pointers. Because of this, therefore, we should not assume that a matrix that is 50% sparse occupies 50% less space. The sparse matrix and ancillary arrays may use only 20% less space or even in some cases, as we shall see, may need more space.

So the first question we should probably ask ourselves is: Do we really need a sparse matrix technique? The author's personal opinion is that if the matrix fits comfortably into the computer's main memory and unless it is very sparse, say 20% or less dense, then we should probably just go ahead and use a two-dimensional array, zeros and all. But if the matrix starts butting against the limitation of main memory, sparse techniques should definitely be used. For computers without virtual memory the sparse version may fit into main memory, obviating the need for backup storage on disk or tape or designing an overlay program. Even for virtual memory systems the sparse version should be faster than a nonsparse program. Paging algorithms are complex, but transparent to the user, who is sometimes tempted to merely dimension his matrices as large as he wants them and not worry about the efficiency of his algorithm. Sparse matrix techniques shove some of the complexity onto the user, but the result will be more efficient matrix operations.

We shall discuss four techniques for manipulating matrices with the zero elements in random locations: the bit map, paired vectors, a linked list, and hashing. The first two use arrays as the basic data structure and may be used easily with FORTRAN, although C offers certain advantages with the bit map. A linked list is by definition a dynamic data structure and cannot be implemented in a language like FORTRAN that offers only the static array as the basic data structure. But C easily handles linked lists. Hashing also employs the array as the basic data structure, although in a somewhat unique fashion. Either C or FORTRAN may be used with hashing, but C offers certain advantages. Nevertheless, our example for hashing will be in FORTRAN which, for better or worse, is the default language of scientific users (and all too often the only one they know).

3.3.1. The Bit Map

With the bit map representation the nonzero element of the matrix are stored in a vector, in either column or row order, and a second area of main memory is established that contains a series of bits arranged in some order. If a bit is one (the bit is set) the corresponding matrix element is nonzero and is present in the array. On the other hand, if the bit is zero (clear) the matrix element is zero and is not present. For example, the sparse matrix

$$\begin{pmatrix} 1 & 0 & 0 & 0 & 8 \\ 0 & 3 & 0 & 0 & 9 \\ 2 & 4 & 5 & 7 & 0 \\ 0 & 0 & 6 & 0 & 0 \end{pmatrix} \tag{3.6}$$

could be stored in column order with its corresponding bit map as

$$\begin{pmatrix} 1 & 0 & 0 & 0 & 1 \\ 0 & 1 & 0 & 0 & 1 \\ 1 & 1 & 1 & 1 & 0 \\ 0 & 0 & 1 & 0 & 0 \end{pmatrix} \begin{bmatrix} 1 \\ 2 \\ 3 \\ 4 \\ 5 \\ 6 \\ 7 \\ 8 \\ 9 \end{bmatrix}. \tag{3.7}$$

The array for the matrix elements offers no representational problems in FORTRAN, but the bit map requires a little thought. Operations on bits are difficult in some programming languages even though the computer's hardware has intructions to manipulate individual bits. But writing a program in assembly language to take advantage of these instructions is an unpleasant prospect. But with some contortions even a language like FORTRAN, not particularly endowed with bit handling features, can be coerced into service. How? Let us see.

As is well known one byte contains eight bits. Suppose that we wish to work with a square matrix of size 256×256 (which looks suspiciously like a power of two). Then the bit map occupies $256^2/8 = 8{,}192$ bytes of memory. This sounds like a lot of memory, but if our matrix is 50% sparse and stored in a double-precision of eight bytes we save 262,144 bytes by not storing it as a two-dimensional array. To be precise our sparse version uses 51.6% of the space needed by a nonsparse two-dimensional array. Space for the matrix array and the bit map is allocated by

BYTE IND (8192)
REAL∗8 A (32768).

In manipulating the bit map we make use of Boolean operators and check for clear and set bits. Some masks with the bit set at the appropriate position within the byte are useful and are defined by

BYTE   MASK(8)
DATA   MASK /x′01′, x′02′, x′04′, x′08′, x′10′, x′20′, x′40′, x′80′/

MASK(1) contains a set bit in the first position within the byte, with the rest clear, MASK(2) a set bit in the second position, and so forth.

Assume that we are entering the nonzero elements in column order. For each matrix element we must also enter the i and j indices that identify the element. If the matrix is of size $m \times n$ the mapping function is $k = m(j - 1) + i$. When the triple i, j, a_{ij} is entered a_{ij} is stored for the first time in A(1) and a counter, initially set to one, is incremented by one. Let the counter be called INDEX. Then k is calculated. Given that FORTRAN integer operations truncate results of divisions, the correct byte with the IND array to set the bit may be found from IBYTE $= (K - 1)/8 + 1$ and the bit within the byte by IBIT $= K - 8*($IBYTE $- 1)$. The derivation of these relations should be straightforward. Then the bit is set with a logical OR operation

$$\text{IND(IBYTE)} = \text{IND(IBYTE)} . \text{OR} . \text{MASK(IBIT)}.$$

Perhaps this needs a little explanation. MASK(IBIT) has zeros in all positions except IBIT, which is one. In the logical OR operation a zero bit in MASK leaves the corresponding bit in IND(IBYTE) alone: a zero remains a zero and a one remains a one. But the one in position IBIT will set the bit in the corresponding position of IND(IBYTE) to one. Then further triples i, j, a_{ij} are entered, a_{ij} going to A(INDEX) and setting the proper bit in the array IND.

After the matrix has been stored along with its bit map the presence or absence of an element may also be found by use of logical operators. If we desire to know if a given a_{ij} is present once more we calculate k and then IBYTE and IBIT. We also need a one byte variable, which may as well be called TEST, to see if IBIT is set within IND(IBYTE). This is accomplished by use of a logical AND operation:

$$\text{TEST} = \text{IND(IBYTE)} . \text{AND} . \text{MASK(IBIT)}$$
$$\text{IF (TEST} . \text{EQ} . \text{MASK(IBIT))} \ldots$$

How does this work? MASK(IBIT) has, again, all bits set to zero except the bit at position IBIT. The AND operation sets all the bits in IND(IBYTE) to zero that correspond to the zero bits in MASK(IBIT). The bit at position IBIT in IND(IBYTE) is left alone. If it happens to be set then TEST is equal to MASK(IBIT). If it is zero then TEST is zero and not equal to MASK(IBIT).

Manipulating bits is easier in C than in FORTRAN because the former offers a data structure, called a "field", that permits access to individual bits. Our array IND could be defined in C as

```
struct {
        unsigned bit 1 : 1;
        unsigned bit 2 : 1;
              ⋮
        unsigned bit 8 : 1;
        } ind[8192];
```

Instead of using masks and logical operators we could assign a value to an individual bit by

$$\text{ind[ibyte].bit 1} = 1;$$

and check for a bit's being set or clear by

$$\text{if (ind[ibyte].bit 5} == 0)$$
$$\vdots$$

Some operations are very easy with sparse matrices stored with bit maps. If the array of elements is stored in column order, then operations with columns present no serious obstacles. Figure 3.2 gives a FORTRAN coding for calculating the Euclidean norm of the columns of a sparse matrix. The program embodies the concepts of bit manipulation in FORTRAN discussed in the text.

But other operations are cumbersome with the bit map. Suppose that our interest is to calculate the Euclidean row norm of a sparse matrix. If the matrix is stored in row order, of course, there is no problem. But if it is stored in column order, then finding the elements in a given row is a headache. The bit map indicates whether a given element is present in a row, but finding the element in the array A is the problem. For a given i, j in a row we would have to scan down the bit map, counting up the nonzero entries, until we reach the element we want. Then we pick a_{ij} from the array A with the count of the nonzero entries. This procedure, obviously, will be extremely slow. Execution time could be improved by maintaining an auxiliary array, which in FORTRAN could be declared INTEGER $*2$, to keep track of how many nonzero elements there are in each column of the sparse matrix. For a 256×256 matrix this only adds an additional 512 bytes of memory. In this way we could, given the j index of the element, sum up the total nonzero elements in the first $j - 1$ columns and start the scan of the bit map from well beyond the beginning of the array. Alternatively, we could transpose the bit map and rearrange the elements of A so that they are in row order, but this involves a lot of data movement.

Unless the operations are to be performed in a way that takes advantage of the order in which the elements of the sparse matrix are stored, such as calculating the Euclidean norm of the columns when the matrix is stored in column order, the bit map cannot be recommended. The paired vectors scheme, discussed next, will be better. For a further discussion of bit maps, see the article by Pooch and Nieder (1973).

3.3.2. Paired Vectors

With paired vectors the nonzero elements are, once again, stored in a vector in either column or row order. Instead of a bit map we use an integer array of the same size as the vector of matrix elements, and whose entries correspond one for one with the entries of the matrix elements and indicate the location of the element within the matrix. Let us refer to the integer array as the index array and the array of the matrix elements as the data array. The matrix (3.6) can be represented as:

Index array	Data array
1	1
3	2
6	3
7	4
11	5
12	6
15	7
17	8
18	9

The index array contains the k index of each element of the data array, $k = m(j - 1) + i$, for an array of size $m \times n$. If the product mn is no greater than 32,767, the index array may be declared INTEGER $*2$ in FORTRAN to save space. (Some languages, like PASCAL or C, allow us to declare an integer without sign. In that case, all sixteen bits of a two byte word may be used to represent numbers from 0 to $2^{16} - 1 = 65,535$. But FORTRAN always works with signed integers; one bit is reserved for the sign permitting the representation of numbers from $-32,768$ to 32,767.) If our matrix is larger then a four byte integer would have to be used. Unless the matrix is very sparse the paired vectors storage scheme for sparse matrices requires more space than the bit map. To be precise, a two byte word contains one k index with paired vectors and the equivalent of sixteen k indices, but for both null and nonzero indices, with the bit map. Therefore, unless the matrix is less than $\frac{1}{16} = 6.25\%$ dense, the bit map is more economical. If we need a four byte word for the k index the cutoff density is 3.125%. Paired vectors, of course, still need less space than the full two-dimensional array if the matrix density is less than 80% for two byte k indices and $66\frac{2}{3}\%$ for four byte indices.

Although paired vectors require more space than the bit map they are, for some operations, more efficient. But before discussing this we should investigate how to locate a given element a_{ij}. First, we calculate the k index of the element and then search for it in the index array. The worst search strategy is the linear search: start at the beginning of the array and look for the index until either it is found or the list is exhausted. The linear search needs an average of $L/2$ comparisons, where L is the size of the array, to find an element or exhaust the list, and is far too time-consuming for large L. Given that the k indices are ordered, by columns for our example, a binary search performs far more efficiently. This standard search technique starts by looking at the middle element of the array. There are three possibilities: the element searched for is equal to the middle element, and the search is over; the search element is greater than the middle element; or the search element is less than the middle element. If the search element is less we discard all the elements from the middle upward, and set the new middle element to the middle of the remaining elements. The same principle applies if the search element is greater than the middle element, except that we discard the lower part of the array. In this way we keep bisecting the remaining part of the array until the search element is found or the array is exhausted. The binary search needs about $\log_2 L$ comparisons, much better performance than the linear search. (An even more efficient search strategy, hashing, will be discussed later.)

Figure 3.3 shows a brief subroutine in FORTRAN to perform a binary search on an array passed to the subroutine as an INTEGER $* 2$ array IX of size L, also a parameter. K is the element desired, and IWHERE is returned as the array location where the element is found, or 0 if the element is not present. The subroutine presents little difficulty of interpretation. Two indicators are used. ILOW is initially set to the array's lower limit of one and IHIGH to the upper limit of L. IMID marks the current middle element of the part of the array that we are examining. It is set to the mean of ILOW and IHIGH. Because IMID is an integer, should the sum of ILOW and IHIGH be odd, it will be truncated. The DO WHILE loop resets the indicators depending on K's comparing high or low with respect to IMID. The loop is executed until either K is found or the list is exhausted.

Even though the binary search is more efficient than the linear search, much time will still be consumed searching if the array is large and we repeat the search for different, random elements. As with the bit map, time will be saved by using an auxiliary array with a count of the number of nonzero elements in each column. Unless the matrix is horrendously large, an INTEGER $* 2$ array suffices. Given an element a_{ij} we know that, if present, it resides in the jth column. The search can start at a position given by the sum of the first $j - 1$ columns plus one and end at the position given by the sum of the first j columns. The auxiliary array requires little additional storage, only 512 bytes for a 256×256 matrix, unless there are many more columns than rows. In that case the matrix should be stored in row, rather than column, order.

In what way does the paired vectors—the name is still appropriate even if we use the auxiliary array—scheme make up for its greater space requirement, compared with the bit map for all but extremely sparse matrices? As already mentioned some operations are more tractable with paired vectors. Row access for a matrix stored in column order, for example, is easier. To illustrate this we shall consider transposing a matrix.

When two-dimensional arrays are used to represent matrices, transposition is simple. Suppose that the matrix is A and of size $m \times n$; its transpose ATRAN is of size $n \times m$. A five line—three lines if the assignment statement is labeled—FORTRAN program suffices:

```
DO J = 1, N
  DO I = 1, M
    ATRANS(J, I) = A(I, J)
  END DO
END DO
```

When paired vectors are used to transpose a sparse matrix we suspect, correctly, that the coding will be more complicated. For matrix (3.6), the paired vectors for the matrix and its transpose are:

Matrix		Transpose	
1	1	1	1
3	2	5	8
6	3	7	3
7	4	10	9
11	5	11	2
12	6	12	4
15	7	13	5
17	8	14	7
18	9	18	6

Our first observation is that the matrix and its transpose need exactly the same amount of space. An in-place transpose is an attractive possibility. But sometimes we wish to keep distinct a matrix and its transpose. Let us start by assigning a separate array for the transpose. The simplest scheme would be to scan down the matrix, picking off the elements and changing the k index to reflect the interchange of i and j. That is, decompose k into its i and j components and then calculate a new k from the mapping function $k = n(i - 1) + j$. We end up with a transpose,

1	1
11	2
7	3
12	4
13	5
18	6
14	7
5	8
10	9

that is no longer in column order, nor is it in row order either.

The potential algorithm could be modified to place the elements in their proper order. After 2 is placed in the second position of the transposed array we pick 3 off the array of the original matrix and find that its k index is lower than that of 2 in the transpose array. So we swap them in both the data array and their indices in the index array. And so on for succeeding elements. This evidently demands a lot of data movement—an unappetizing proposition. Alternatively, the data and index arrays of the transpose could be sorted. But even with a good sorting algorithm, such as quicksort, we will be doing excessive work. How much? Let us consider a concrete example.

To transpose an $m \times n$ matrix using two-dimensional arrays requires $O(mn)$ operations. Assume that we have a square 256×256 matrix that is 50% dense. With two-dimensional arrays, about $256^2 = 65{,}536$ operations would be necessary. The simple algorithm suggested two paragraphs earlier would do $0.5 \cdot 256^2 = 32{,}768$ operations. So far, so good. But a quicksort performs $O(L \log_2 L)$ operations, where L is the size of the data set being sorted. The sort does 491,520 additional operations. Altogether, we need fifteen times as many operations. This is too heavy a burden for the advantage of a simple algorithm.

Fortunately, at the expense of a more complicated algorithm and the additional space of some auxiliary arrays, the sparse matrix may be transposed with $O(mn)$ operations. Figure 3.4 illustrates this. The sparse matrix is stored in column order in A with associated index array IND1, its transpose in column order in QR with index array IND2, and ICOL1, ICOL2, and IL are the auxiliary arrays. ICOL1 and ICOL2 contain the counts of the nonzero elements in the columns of A and QR, respectively, and IL is used as a counter while transposing. The triples i, j, a_{ij} are entered, the k index calculated, and both a_{ij} and k stored in their respective arrays in the order given by the continuous counter INDEX (it is assumed that the triples i, j, a_{ij} are entered in column order). An i index of -999 signals the end of the list.

The heart of the program incorporates a pair of DO loops. The first, DO I = 1, N, is a row counter and the second, DO J = 1, N, a column counter. For a given row and column we skip over the elements in the preceding columns; this is controlled by ISUM. Then for a given column we start initially at the first position within the column IL(J) = 1. A subroutine DECOMIJ breaks the k index into its i and j components. (For the most efficient operation it would be best to use in-line code rather than a subroutine call. It is ironic that the pundits of good programming style invariably recommend breaking a program into modules, but in practice at the first sign of inefficiency these rules go by the board. What, then, is good programming "style"?) If the i components match, we have the correct nonzero element, which is placed in the array QR at a position given by the continuous counter INDEX, used previously for filling in the array A. A new k index is computed as $k = n(i-1) + j$, to indicate the correct position of the transposed element. The continuous counter INDEX and the starting position within the column counter IL(J) are incremented by one and we go on to the next element within the column. When the column is exhausted the number of nonzero elements within the column of the transposed matrix is calculated, and we go on to the next row.

Unless the transposed matrix is needed in its own right, the space allocated for it can be saved by doing an in-place transpose by use of the space already allocated for the original matrix. Previous considerations should lead us to think that an in-place transpose will likely involve much data movement or excessive computing time. Nor will our expectations be unfulfilled. A glance at the paired vectors representation of matrix (3.6) and its transpose shows that the first element retains its position in the transpose. The 8 is moved to the second position, necessitating a movement of the other elements downward. And so forth. The indexing scheme, to keep track of where the elements are to go, is likely to be complicated. As an alternative, change the k indices in the index array to reflect swapping the i and j indices, and do a quicksort which involves us with an $O(L \log_2 L)$ algorithm. Too slow. (It would be slower yet if we were to use a sorting algorithm like bubblesort, whose performance is $O(L^2)$. Then we could spend a year's computing budget just transposing a large, sparse matrix.)

But once again there is, fortunately, an easier way. This easier way depends upon the observation that, in Figure 3.5, the variable ISTART locates the needed element in the array A that is to be stored in the transpose QR. If, instead of physically moving the element, we store the value ISTART in an array—call it NEXT—then a reference to A(NEXT(ISTART)) is really a reference to the transposed element. No data movement at all is necessary. What is more, the space assigned for the array IND2, which is no longer needed, can be reassigned to the array NEXT. Likewise, the k index found in IND1(NEXT(ISTART)) can be used. After decomposition into i and j, these must be interchanged.

All of this will be clearer if illustrated in a program—Figure 3.5, which closely parallels the coding of Figure 3.4. There are no arrays QR or IND2; the NEXT array takes the place of IND2. To print out the transpose we merely use IND1 (NEXT(L)) instead of IND1 (L) inside of the DO $L = 1,2000$ loop. The call to DECOMIJ within the loop reverses the parameters i and j rendering unnecessary their subsequent interchange.

The algorithm for an in-place transpose requires space for both the IND1 and the NEXT arrays. If the product mn is less than 32,767, INTEGER $*2$ arrays are sufficient. Otherwise, they must be declared INTEGER $*4$. For the former the matrix should be at most $66\frac{2}{3}\%$ dense to make this scheme worthwhile, and for the latter at most 50% dense.

3.3.3. The Linked List

So far, in the discussion of sparse matrix techniques, the array has been the only data structure used. But we have also left unaddressed the question of fill-in, except where it did not occur, as with inverting triangular matrices, or where it was perfectly predictable, as with the backwards product of two vectors. Unfortunately, in the general case of operations with sparse matrices, the amount of fill-in can be difficult to ascertain before-hand and it can also be highly variable. At the one extreme, no fill-in at all occurs as with

$$\begin{pmatrix} 1 & 0 & 0 \\ 0 & 2 & 0 \\ 0 & 0 & 3 \end{pmatrix} \begin{pmatrix} 0 & 0 & 4 \\ 0 & 5 & 0 \\ 6 & 0 & 0 \end{pmatrix} = \begin{pmatrix} 0 & 0 & 4 \\ 0 & 10 & 0 \\ 18 & 0 & 0 \end{pmatrix}. \tag{3.8}$$

At the other extreme, we have something like

$$
\begin{pmatrix} 1 & 0 & 0 \\ 2 & 0 & 0 \\ 3 & 0 & 0 \end{pmatrix} \begin{pmatrix} 4 & 5 & 6 \\ 0 & 0 & 0 \\ 0 & 0 & 0 \end{pmatrix} = \begin{pmatrix} 4 & 5 & 6 \\ 8 & 10 & 12 \\ 12 & 15 & 18 \end{pmatrix}, \tag{3.9}
$$

where the product of two matrices, each one of which is only $33\frac{1}{3}\%$ dense, results in a matrix that is 100% dense.

One way out of the conundrum, and in an ideal world the best, is to analyze the problem sufficiently so as to know how much fill-in will occur. Thus, with Eq. (3.8), we have the product of an antidiagonal matrix premultiplied by a scaling matrix. No fill-in will take place. In Eq. (3.9) each of the matrices to be multiplied is subrank, rank one to be explicit. The product matrix is also rank one and really need not be stored as a full 3×3 matrix. It would be sufficient to store one of the rows as a vector and also to store a vector of the coefficients to multiply this row. For both of these examples we could readily use arrays and dimension them according to the conclusions of our analysis regarding fill-in.

But suppose predicting the amount of fill-in surpasses our analytical powers—not everyone is a Gauss—or that the amount of fill-in cannot be predicted? We could go ahead and assign some dimension to the array, such as the array for the product of two matrices, and hope for the best, but running the risk of either grossly overdimensioning the array or having execution fail because of array overflow. A better solution is to employ a data structure other than the static array, a dynamic data structure.

There are many dynamic data structures, which go by names such as stacks, queues, lists, trees, B-trees, dequeues, and are characterized by being able to grow as we put elements into them and shrink as we take elements out. This dynamism offers, for certain applications, a distinct advantage over the static array. (Static array because some programming languages, such as PL/I, permit dynamic arrays whose size is specified at execution-time, rather than at compile-time as with FORTRAN. The space allocated for the array can also be freed and reallocated during execution of the program. The dynamic array, although useful, does not solve the problem of fill-in because whether at compile- or execution-time an array size must be known. PL/I does, however, support the other dynamic data structures.) Sparse matrix techniques with dynamic data structures, although feasible and even desirable when fill-in is difficult to predict, are little used, largely, we suspect, because the perennially popular scientific programming language FORTRAN does not support them. They do, however, suffer some drawbacks, as we shall see. Many books discuss dynamic data structures, of which Wirth (1976) and Horowitz and Sahni (1984) can be recommended.

One of the simplest dynamic data structures is the unidirectional linked list. Unlike the array, where the elements are located in consecutive memory locations, with a linked list the elements may be located nonconsecutively anywhere in memory. Because of the nonconsecutiveness something is necessary to link the elements together. This "something" is a pointer. Pointers are familiar to those who program in languages such as C and PL/I, but somewhat of a mystery to most FORTRAN programmers. Basically, a pointer is a variable, generally declared an integer, that contains instead of a value an address. In Figure 3.5 the NEXT array may be considered an array of pointers, because the array elements themselves merely point to the elements in the data we want to access. This is evident in an expression like IND1(NEXT(L)). The unidirectional linked list may be imagined schematically as

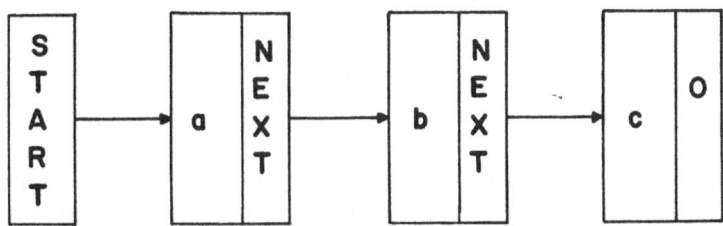

START is a special pointer that contains the address of the first element of the list, a. Each list element contains a pointer to the next element, NEXT, and the last list element, c, has its pointer set to a special value, zero in this instance.

For the list to be truly flexible, the programming language that implements it should incorporate a dynamic memory allocation feature. That is, we imagine the list initially as consisting of only the pointer START, containing zero. An element a is read in, perhaps via the terminal, and at this moment space in memory is assigned for it. The address of this memory is assigned to START and NEXT of a set to zero. When b is read in the memory allocated for it is placed in the NEXT field of a; NEXT of b is set to zero. And so forth. Without dynamic memory allocation we could form a list by allocating an array large enough to contain all the list entries and letting the list elements occupy varying positions within the array. This is, in fact, a scheme sometimes used by FORTRAN programmers to implement lists. But we really should not consider such a scheme a genuine list. With a few list elements we waste memory because of an overdimensioned array, and with many elements we run the risk of array overflow. A real lists allocates only as much space as needed and will never overflow until memory is exhausted.

Because the list elements occupy noncontiguous memory locations, insertion and deletion of elements is easy. A new element—call it d—may be added anywhere. Suppose that we wish to insert it between a and b. Space is allocated for d, the addresss of the space is assigned to the NEXT pointer of a, and the NEXT pointer of d is set to the address of b:

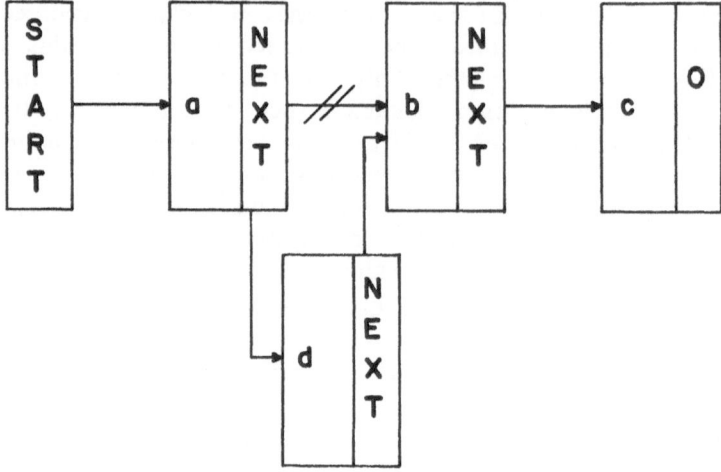

Do we wish to delete the element *b*? Simple. Let the NEXT pointer of *a* point to the address of *c*. Nothing now points to *b*, and the space that it occupies will sooner or later be overwritten.

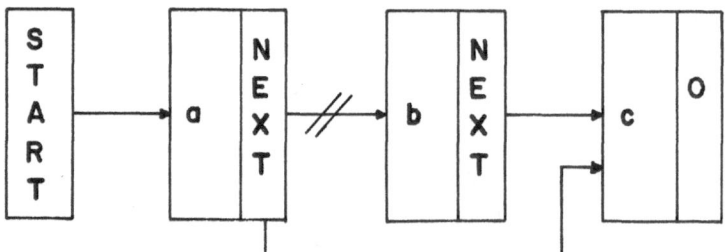

But what has all of this to do with sparse matrices? If a linked list is used to represent a sparse matrix, fill-in, should it happen, offers no difficulty: the fill-in elements are inserted into the list in their proper position. "Proper" is determined by the order we desire for the matrix elements in the list, perhaps column order.

Many programming languages permit dynamic data structures; PL/I has already been alluded to. C is an increasingly popular language with many computer users and one that incorporates dynamic data structures. We shall use it to present, in Figure 3.6, a linked list implementation of a program for transposing a matrix. The program does not, in fact, illustrate the advantages of a linked list for handling the problem of fill-in because, with matrix transposition, fill-in does not take place. An example, by its nature, cannot be too complicated, and it is more important at the moment to show how a linked list can be implemented and manipulated in a language like C, because the concepts involved are likely to be foreign to the average FORTRAN programmer. Later on, some comments will be made concerning more general applications of linked lists. (Actually, FORTRAN-77 and DEC's VMS operating system allow us to build a genuine linked list. VMS provides two routines from the Run Time Library, LIB$GET-VM and LIB$MO VC3, that are callable from FORTRAN. The former allocates virtual memory dynamically, and the latter moves values to the memory allocated. But this is a pleasing characteristic of the VMS operating system, not an inherent virtue of FORTRAN—and unavailable on most operating systems.)

For the benefit of readers unfamiliar with C Figure 3.6 will be explained in some detail. C makes great use of pointers, identified by preceeding the variable by an asterisk as with ∗iptr. The function ∗malloc ( ) allocates dynamic memory, the address of which can be assigned to a pointer. A statement such as iptr = malloc (16) allocates sixteen bytes of dynamic memory and places the address of the beginning of the block of sixteen bytes in the pointer variable iptr. Another convenient feature of C, the structure, is lacking in FORTRAN: it permits us to group different data types together under one name. The definition

```
struct
{
    short int      ind1;
    short int      ind2;
    double         aij;
    long int      ∗ ichain;
} matrix;
```

groups two short integers (two bytes on most systems), a double-precision floating-point variable, and a long integer (four bytes on most systems), together as the structure "matrix". Individual structure members are indentified by the structure name followed by the member name with an intervening decimal point, matrix . aij, for example. C also uses preprocessor statements, identified by #. A statement such as # define NULL 0, merely substitutes a zero for NULL every time the latter is encountered and before the program is compiled. A C program starts execution with the statement, main ( ). Preprocessor statements should precede main ( ).

In Figure 3.6 the structure definition "matrix" occurs before main ( ). After main ( ) we encounter a statement, struct matrix ∗ mat. What is happening is that the structure "matrix" is being considered a template. No space is actually reserved for it until the definition of ∗ mat within the main program is encountered. That definition states that ∗ mat is a pointer to a structure of the type "matrix". Within the program we could have, and in a less trivial application would have, other structure pointers, for example, ∗ mat1 and ∗ mat2, to the same type of structure. That is, a statement such as struct matrix ∗ mat1, ∗ mat2 allocates space for two structures, each of the type matrix—two short integers, one double-precision variable, one long integer pointer. The addresses of the memory allocated will be assigned to the pointers ∗ mat1 and

∗ mat2. In the program itself there is no structure "matrix." The program of Figure 3.6 defines some more variables—in a C program *all* variables must be declared, there are no defaults—including *aij*. Is this not a double definition of that variable? No. One *aij* occurs within the structure template "matrix" and in the program itself the full name for this variable is matrix.aij, completely distinct from *aij*.

The heart of the program starts with "while". Every time a triple of the variables *i, j* and *aij* is read in, sixteen bytes of dynamic memory are allocated and the address of the memory is assigned to the pointer ∗ iptr. The structure pointer ∗ mat is also assigned this address. A sentence like mat → ind1 = *i* takes the value for *i* just read in and transfers this value to the memory location assigned for matrix.ind1. Let us be clear about what is going on. malloc (16) requests sixteen bytes of memory and assigns the address of the beginning of the block of sixteen bytes to both ∗ iptr and ∗ mat. Assume that the block starts at location 16,000. This address will be assigned to the pointer ∗ iptr and to ∗ mat. Because ∗ mat is a pointer to a structure of the type "matrix", the program assumes that memory locations 16,000 and 16,001 will be filled with a short integer, as will locations 16,002 and 16,003. An eight byte double-precision number fills locations 16,004 through 16,011. A four byte long integer, assumed to contain an address, occupies locations 16,012 through 16,015. The various scanf statements, similar to READ in FORTRAN, read in values for *i, j*, and *aij*, storing them somewhere in memory. Assume that they go to memory locations 2,000, 3,000, and 4,000. The statement mat → ind1 = *i* moves the two bytes at location 2,000 to the two bytes starting at 16,000.

The first time we read in values the pointer ∗ start, indicating the beginning of the list, is set to the memory location with the first matrix element and its indices. To form the linked list some way is needed to set the ∗ichain pointer to contain the address of the area where the next element and indices will be stored after they are read in. This is done by use of the ∗ oldptr pointer. When values are read in and ∗ iptr set, the pointer ∗ mat is set to the address of ∗ oldptr, which contains the previous, rather than the current, value of ∗ iptr. The ∗ ichain pointer is set to the current ∗ iptr. When the last matrix element is read in, ∗ ichain is set to NULL to signal the end of the list.

To leaf through the list we set the ∗ mat pointer to ∗ start at first and to ∗ ichain on subsequent reads until NULL is encountered. The matrix is transposed by interchanging the *i* and *j* indices. If the matrix has been kept in column order, the transpose will be in row order.

The integers for the *i* and *j* indices in the structure template, ind1 and ind2, may be short integers because it is unlikely that they will exceed 32,767. We could also have used a mapping function to define a *k* index, but as *k* would have to be a long integer nothing would be saved. The ∗ ichain pointer should be a long integer because it contains an address of a memory location that will most likely be greater than 32,767. The structure for the linked list requires double the space of that needed by an array. Our matrix should, therefore, be at least 50% sparse before the linked list is competitive.

The unidirectional linked list suffers a serious drawback. Finding an individual element is time-consuming. If the elements are stored in random order, the entire list must be searched for the element. This inconvenience can be alleviated somewhat by storing the elements in some order, such as column order. For an operation such as matrix multiplication, one matrix will be stored in column order and the other in row or column order.

To achieve efficiency with the linked list we must increase the complexity of the list. Pooch and Nieder (1973) discuss one palliative measure, keeping the matrix in order and maintaining a table of pointers to groups of elements. Horowitz and Sahni (1984) present an algorithm in PASCAL to implement a circularly linked list. Such a list includes more pointers, pointers to the next element in the same row, the next element in the same column, and others. Both of these schemes demand even more space. Our matrix should be even more sparse than 50%, considerably more sparse.

As another alternative we could employ trees. Trees, particularly binary trees, are much used in computer science and are, therefore, discussed in books on computer data structures. A binary tree (B-tree) representation of a sparse matrix with a total of L nonzero elements permits access of a random element in $O(\log_2 L)$ searches, superior to the $O(L)$ for the unidirectional list.

3.3.4. Hashing

The tree is a data structure that grows and shrinks with the elements added and deleted. At the cost of decreased flexibility we can opt for a data structure with better search time than the $O(\log_2 L)$ of the tree. If the maximum number of elements in the sparse matrix is known or can be reliably estimated, a properly designed hash table allows retrieval of an element after one or two, possibly three, tries. An unsuccessful search—the element is not present—may take slightly longer, perhaps five or six tries, but is still not excessively time-consuming. Although hashing may be implemented by a dynamic data structure, it is more common to use an array, to the great relief of FORTRAN programmers.

To briefly explain hashing remember what is done when a two-dimensional array represents a matrix. In memory, the array is laid out linearly by use of a mapping function that transforms the two indices i and j to one index k. FORTRAN maps the array so that columns are stored consecutively. C and PL/I store rows consecutively. Imagine our array laid out linearly with the k index next to each data element a_{ij}. If the matrix is sparse, many of the a_{ij} are zero. For each zero a_{ij} we could set its k index to zero also, as the element contributes nothing to the information content of the matrix and we have no interest in where it is. Only the nonzero elements interest us. If the list of the k and a_{ij} pairs could be compressed, by omission of the zero pairs, space would be saved.

Hashing, at least partially, solves the problem by transforming k to a smaller range. Suppose that the matrix is $1{,}000 \times 1{,}000$ and is 90% sparse. k ranges from 1 to 10^6, but only 10% of the available k, a_{ij} pairs are nonzero. If we could transform the k index to run from 1 to 10^5 a smaller array would be sufficient. Also, once k is given we could immediately retrieve the corresponding a_{ij} by going to the kth location in the list, which we assume to be held in an array. A difficulty of hashing arises from the nature of the transformation. We shall from now on refer to the k indices as keys, which is standard hashing terminology. Because the number of keys is greater than the number of array locations, the transformation is many to one. Consider transforming 10^6 k indices to an array of size 10^5. If our key transformation is simply $k' = k/10$ we indeed compress the range of k', but all k indices from 1–9 are transformed to 0, indices 10–19 to 1, and so forth. Two distinct keys transforming to the same array location is known as a collision. A way must be found to resolve collisions and assign unique array locations to the unique keys.

Because our main interest is using hashing to represent sparse matrices, we will not go into great detail. The reader may refer to a good book on data structures. It suffices to say that the array size should be $b = 4j + 3$ and j a prime number. Such a choice for b, if the transformed keys are calculated as $k' = k$ modulo b, distributes the transformed keys uniformly over the array and minimizes collisions. The hash table, namely the array, should be at least 20% larger than the number of keys to be hashed for efficient run-time performance. Standard hashing terminology refers to a loading factor α, just the number of keys divided by the array size. We never want $\alpha = 1$ (number of keys equal to array size) and prefer $\alpha \leq 0.8$ or even lower.

Of the many mechanisms for resolving collisions the easiest to understand conceptually is linear probing. To understand how it works let us be given eight numbers in the range 1–1,000. The numbers, as found from a random number generator are: 28, 164, 228, 402, 595, 679, 787, and 939. Take for the size of the array $b = 11$ ($4j + 3$ with $j = 2$). To allow for some of the hashed value's possibly being zero, the array will be numbered from 0 to 10. The first number, 28, divided modulo 11 hashes to 6. Our array initially looks like

0	1	2	3	4	5	6	7	8	9	10
						28				

164 hashes to 10 and 228 to 8. So far, so good. Our array looks like

0	1	2	3	4	5	6	7	8	9	10
						28		228		164

But 402 hashes to 6, already occupied. What do we do? With linear probing we start at location 6 and begin searching upwards for the first available empty location, called bucket in hashing terminology. The collided value is deposited there. Following this philosophy we deposit 402 in bucket 7. 595 hashes to 1, no problem. But 679 collides with 228, already in bucket 8, and hence goes into bucket 9. 787 again hashes to 6, the third collision for this number, and buckets 6, 7, 8, 9, and 10 are already occupied. Therefore, we wrap around the array and start at bucket 0, unoccupied and thus 787 goes there. 939 hashes to 4 and is deposited in that bucket. At the end, the array looks like

0	1	2	3	4	5	6	7	8	9	10
787	595			939		28	402	228	679	164

To find a number we first hash it to obtain the transformed key and go directly to the bucket with the transformed key. If the content is the same as the number the search is over. If not, we start a linear search until either the number is found or an empty bucket encountered, indicating that the number is not present. Given 28 we hash it to 6 and find it immediately at that bucket. Had we started with 787 the hashed key of 6 would result in a collision with the 28 already in that bucket. We start a linear search that bumps against the array upper limit at 10. Wrapping around we start at 0 where 787 is found after a total of six comparisons. The number 679 would be found after two comparisons. Given 39 we hash it to 6 to find an occupied bucket and start a linear search that ends at the empty bucket at 2. Eight comparisons and we know that 39 is not present.

D. Knuth (1973) demonstrates, in his *The Art of Computer Programming: Sorting and Searching*, that with linear probing and a loading factor α a successful search needs an average of

$$S \cong \tfrac{1}{2}\left(1 + \frac{1}{1 - \alpha}\right)$$

comparisons and an unsuccessful search an average of

$$U \cong \tfrac{1}{2}\left[1 + \frac{1}{(1 - \alpha)^2}\right]$$

comparisons. With $\alpha = 0.5$ and with the hash table half full, 1.5 comparisons suffice for a successful search and 2.5 for an unsuccessful one. Both S and U climb rapidly as α increases. For $\alpha = 0.8$, we calculate $S = 3$ and $U = 13$; for $\alpha = 0.9$, $S = 5.5$ and $U = 50.5$. It appears that for linear probing α should be even less than 0.8, perhaps $\alpha \leq 0.5$. Other methods for resolving collisions, not discussed here, do better. With quadratic probing and $\alpha = 0.8$, for example, we find $S = 2$ and $U = 5$. A serious sparse matrix program that relies on hashing should probably use one of the more advanced methods for resolving collisions.

Figure 3.7 gives FORTRAN coding to calculate the Euclidean row norm of a sparse matrix stored in column order. Hashing is used to locate the elements required to sum their squares in a given row. The program could be written somewhat more elegantly in C. We could use a structure array, rather than paired arrays, to form the hash table with the data elements and corresponding index. But at least FORTRAN is capable of implementing hashing, unlike its poor showing with linked lists.

How much space do we need with a hash table? The index array of Figure 3.7 uses four bytes for each k index and eight bytes for each data element, twelve bytes altogether. With a two-dimensional array we need eight bytes per data element, or a total of $8mn$ bytes for an $m \times n$ matrix. With hashing we need $12mn$ times the denseness of the matrix divided by the loading factor. If s is the denseness, measured from 0 to 1, we have $12mns/\alpha$. If $\alpha = 0.8$ our matrix should be no more than 53%, for hashing to be attractive. For $\alpha = 0.5$, as with linear probing, then the denseness should be no more than 33%.

Although hashing permits rapid access to random matrix elements, like all methods it has its disadvantages. The most glaring is requiring good knowledge of the sparseness of the matrix and the possible amount of fill-in. Otherwise, we run the risk of either poor storage use because α is too small or poor performance, perhaps even table overflow, if α approaches or exceeds 1. If our programming language incorporates a memory allocation feature, like the malloc ( ) function of C, it is possible to dynamically increase the size of the hash table at run time should it start to fill up. We could allocate more space for a new table, transfer the indices to it, and deallocate the space for the old table. But this is bound to be time-consuming because of the need to recalculate the hash keys; they will not be the same for both tables.

Another drawback is manifest if we decide to delete entries in a hash table. Deletions are cumbersome. Insertions present no problem as long as the table is large enough, but deletions require more than just setting the entry to zero or some other indicator of an empty bucket. We lose the connection to other entries that may have hashed to that bucket. The problem can be handled by using a special sentinel to indicate that a value was deleted. Other keys that hash to the same bucket can still use it as their reference point for probing. Nevertheless, the sentinel itself and the associated data element take up space. With many deletions space will be wasted, exactly what we want to avoid. Unless the conditions for the efficient use of hashing are met (the creation of new elements, but still leaving the resulting matrix sparse, and the deletion of a few old elements), the method will present more problems than advantages.

3.4. Conclusions

Manipulation of a sparse matrix that involves the creation of many new elements, the deletion of old ones, or even column or row interchanges, is a complex undertaking. None of the methods discussed so far is adequate for the task with the exception of trees and circularly linked lists, mentioned briefly in Section 3.3.3. Just interchanging columns will tax methods like the bit map and paired vectors unless something like the NEXT auxiliary array is used. (This is one of the reasons why George and Liu (1981) restrict themselves to positive definite systems. The decomposition of the system by Cholesky's method, discussed in Chapter 5, can be performed without row or column interchanges. Because they use the perennially popular but sometimes inadequate language FORTRAN they realize a considerable saving of effort.)

But if column interchanges are combined with element creation the task becomes hopeless unless a dynamic data structure is used. Even the manipulation of a single matrix can entail the creation of new elements. In obtaining a solution in the L_1 norm of an overdetermined system, discussed in Chapter 6, the reduction of the matrix of coefficients includes a loop where a multiple of one of the columns is subtracted from another, something like $a_{ij} = a_{ij} - da_{kj}$ for given i, and k and j running from one limit to another. i may be a sparse column, but if k is a dense column the transformed i column will also be dense; elements are created. The L_1 algorithm also does extensive column interchanging, largely to enhance stability. For an L_1 algorithm for sparse matrices a language like FORTRAN is hopelessly inadequate. It could, perhaps, be coerced into duty, but a language like C with dynamic data structures is far more preferable. The author emphasizes this point because some scientists feel that FORTRAN is a ideal language for all of their programming needs. It is not.

In conclusion, sparse matrix techniques are many and varied. They are more complicated, considerably so, than simply using a two-dimensional array, and because of the space needed by ancillary arrays or pointers should only be employed if the matrix is very sparse, at most 66% dense: even less dense for some of the techniques. Dynamic data structures, not available in the *de facto* standard scientific programming language FORTRAN, are ideal for some sparse matrix applications. Finally, sparse matrix techniques are a tool to be used when applicable. We should avoid the tendency, sometimes almost irresistible, to become so enamored of them as to give them undue importance. If our problem is amenable to treatment by a two-dimensional array, use a two-dimensional array.

```
- C
  C    IN - PLACE INVERSION OF A TRIANGULAR MATRIX
  C
                  PROGRAM TRI–INV
                  IMPLICIT REAL * 8 (A - H, O - Z)
                  DIMENSION A ( 55 )
  C    READ IN VALUE FOR N AND MATRIX ELEMENTS
                  TYPE *, 'WHAT IS N?'
                  ACCEPT *, N
                  DO J = 1, N
                    DO I = 1, J
                       TYPE *, 'A ( ', I, ', ', J, ') = ?'
                       ACCEPT *, A (J * (J - 1) / 2 + I)
                    END DO
                  END DO
                  WRITE ( *, 5) ((I, J, A (J * (J - 1) / 2 + I), I = 1, J), J = 1, N)
                  TYPE *, '
  C    MAIN LOOP FOR CALCULATING INVERSE ELEMENTS
                  DO I = 1, N
                     A (I * (I - 1) / 2 + I) = 1.0 / A (I * (I - 1) / 2 + I)
                  END DO
                  DO I = 1, N - 1
                    DO J = I + 1, N
                      SUM = 0.0
                      DO L = 1, J - 1
                        SUM = SUM + A (L * (L - 1) / 2 + I) * A (J * (J - 1) / 2 + L)
                      END DO
                      A (J * (J - 1) / 2 + I) = -A (J * (J - 1) / 2 + J) * SUM
                    END DO
                  END DO
  C    PRINT OUT RESULTS
                  WRITE ( *, 5)  ((I, J, A (J * (J - 1) / 2 + I), I = J, N), J = 1, N)
                  FORMAT (2 ( ' ', I2),  ' ', D22.15)
                  END
```

Figure 3.1. In-place inversion of upper triangular matrix.

```
C
C    CALCULATE THE EUCLIDEAN NORM OF THE COLUMNS OF A SPARSE MATRIX
C    STORED AS A BIT MAP
C
             PROGRAM BITMAP
             IMPLICIT REAL * 8 (A - H, O - Z)
             REAL * 8  A ( 32768 )
             BYTE IND ( 8192 ), MASK ( 8 )
             BYTE TEST
             DATA MASK / '01 'X, 02 'X, '04 'X, '08 'X, '10 'X, '20 'X, '40 'X, '80 X /
C    READ IN INPUT VALUES ; I INDEX OF -999 INDICATES END OF LIST
             INDEX = 1
             TYPE * , 'WHAT IS M ? '
             ACCEPT * , M
             TYPE * , 'WHAT IS N ? '
             ACCEPT * , N
1            TYPE * , 'I = ? '
             ACCEPT * , I
             IF ( I. EQ . -999 ) GOTO 5
             TYPE * , 'J = ? '
             ACCEPT * , J
             TYPE * , 'AIJ = ? '
             ACCEPT * , AIJ
C    FORM BIT MAP IN ARRAY IND . STORE MATRIX ELEMENTS IN A
             K = M * ( J - I ) + I
             IBYTE = ( K - 1 ) / 8 + 1
             IBIT = K - 8 * ( IBYTE - 1 )
             IND ( IBYTE ) = IND ( IBYTE ) . OR . MASK ( IBIT )
             A ( INDEX ) = AIJ
             INDEX = INDEX + 1
             GOTO 1
C    CALCULATE EUCLIDEAN NORM OF COLUMNS
5            INDEX = 1
             DO J = 1, N
                ENORM = 0.0
                DO I = 1, M
                   K = M * ( J - 1 ) + I
                   IBYTE = ( K - 1 ) / 8 + 1
                   IBIT = K - 8 * ( IBYTE - 1 )
C    TEST TO SEE IF I, J ELEMENT  IS PRESENT IN ARRAY A
                   TEST = IND ( IBYTE ) . AND . MASK ( IBIT )
                   IF ( TEST . EQ . MASK ( IBIT )) THEN
                       ENORM = ENORM + A ( INDEX ) ** 2
                       INDEX = INDEX + 1
                   END IF
                END DO
                ENORM = SQRT ( ENORM )
                TYPE * , 'THE EUCLIDEAN NORM OF COLUMN ', J , 'IS ', ENORM
             END DO
             END
```

Figure 3.2. Calculation of Euclidean column norms of sparse matrix stored as bit map.

```
SUBROUTINE  BINSRCH  ( IX , L , IWHERE , K )
INTEGER  *  2   IX ( 1 )
ILOW  =  1
IHIGH  =  L
IWHERE  =  0
DO WHILE  ( ILOW .LE. IHIGH )
   IMID  =  ( ILOW  +  IHIGH ) / 2
   IF  ( K .EQ. IX ( IMID ) )  THEN
      IWHERE  =  IMID
      RETURN
   ELSE
      IF  ( K .GT . IX ( IMID ) )  ILOW  =  IMID + 1
      IF  ( K .LT. IX ( IMID ) )  IHIGH  =  IMID - 1
   END  IF
END  DO
RETURN
END
```

Figure 3.3. Binary search routine.

```
C
C     TRANSPOSE A SPARSE MATRIX STORED AS THE VECTOR PAIR  A -  IND1.  A  CONTAINS THE
C     NON-ZERO ELEMENTS AND IND1 THE  K  INDEX OF THE NON-ZERO ELEMENT.  A  IS STORED
C     IN COLUMN MAJOR ORDER. IF  A  IS A  M  BY  N  MATRIX THEN  K  =  M * ( J - 1 )  +  I.  QR  IS
C     THE TRANSPOSE OF  A  WITH IND2 ITS CORRESPONDING INDEX ARRAY.  IL IS AN
C     AUXILIARY ARRAY THAT AIDS IN LOCATING THE DESIRED ELEMENT IN  A  COLUMN OF  A.
C
         PROGRAM TRANS
         IMPLICIT  REAL *8 ( A - H , O - Z )
         INTEGER  *2  IND1 (2000) , IND2 (2000) , ICOL1 (200) , ICOL2 (200) , IL (200)
         DIMENSION  A (2000) , QR (2000)
         TYPE  * ,  'WHAT IS N ? '
         ACCEPT  * , N
         TYPE  * ,  'WHAT IS M ? '
         ACCEPT  * , M
C
C     INITIALIZE .
C
         INDEX  =  1
         DO  J  =  1 , N
            IL ( J )  =  1
         END  DO
         DO  I  =  1, 20
            QR ( I )  =  0.0
         END  DO
1        TYPE  * ,  'I  =  ? '
         ACCEPT  * , I
C
C     USE -999 TO SIGNAL THE END OF THE LIST.
C
         IF ( I . EQ . -999 )  GOTO 5
         TYPE  * ,  'J  =  ? '
         ACCEPT  * , J
         TYPE  * ,  'A ( I , J )  =  ? '
         ACCEPT  * , AIJ
```

Figure 3.4. Transpose of sparse matrix stored as paired vectors.

```fortran
              K = M * ( J - 1 ) + I
              A ( INDEX ) = AIJ
              IND1 ( INDEX ) = K
              INDEX = INDEX + 1
              ICOL1 ( J ) = ICOL1 ( J ) + 1
              GOTO 1
5             INDEX = 1
              DO I = 1 , M
                INDEXHOLD = INDEX
                ISUM = 0
                DO J = 1 , N
                  IF ( ICOL1 ( J ) . NE . 0 ) THEN
                    IF ( J . GT . 1 ) ISUM = ISUM + ICOL1 ( J - 1 )
                    ISTART = ISUM + IL ( J )
                    IF ( ISTART . EQ . ( ISUM + ICOL1 ( J ) + 1 ) ) GOTO 10
                    K = IND1 ( ISTART )
                    CALL DECOMIJ ( K, M, N, I I, JJ )
                    IF ( I I . EQ . I ) THEN
                      K = N * ( I I - 1 ) + JJ
                      IND2 ( INDEX ) = K
                      QR ( INDEX ) = A ( ISTART )
                      INDEX = INDEX + 1
                      IL ( J ) = IL ( J ) + 1
                    END IF
                  END IF
10                CONTINUE
                END DO
                ICOL2 ( I ) = INDEX - INDEXHOLD
              END DO
C
C     PRINT THE TRANSPOSED MATRIX.
C
              DO L = 1 , 20000
                IF ( IND2 ( L ) . EQ . 0 ) GOTO 15
                K = IND2 ( L )
                CALL DECOMIJ ( K, N, M, I, J )
                WRITE ( * , 20 ) I, J, QR ( L )
20              FORMAT ( 2 ( ' ' , I 3 ) , '      ' , D22.15 )
              END DO
15            CONTINUE
              END
C
C     THIS SUBROUTINE DECOMPOSES K BACK INTO ITS COMPONENT I AND J PARTS.
C
              SUBROUTINE DECOMIJ ( K, M, N, I, J )
              J = ( K - 1 ) / M + 1
              I = K - M * ( J - 1 )
              RETURN
              END
```

Figure 3.4 (*continued*)

```
PROGRAM  INTRANS
IMPLICIT REAL * 8  ( A - H , O - Z )
INTEGER * 2  ICOL1 ( 21365 ) , ICOL2 ( 21365 ) , IL ( 21365 )
INTEGER * 4  IND1 ( 363205 ) , NEXT ( 363205 )
DIMENSION  A ( 363205 )
N = 17
M = 21365
INDEX = 1
DO  J = 1, N
  IL ( J ) = 1
END  DO
DO  J = 1, 17
  DO  I = 1, 21365
      NEXT ( I ) = I
      K = M * ( J - 1 ) + I
      A ( INDEX ) = I + J
      IND1 ( INDEX ) = K
      INDEX = INDEX + 1
      ICOL1 ( J ) = ICOL1 ( J ) + 1
  END  DO
END  DO
INDEX = 1
CALL  LIB$INIT-TIMER
DO  I = 1, M
  INDEXHOLD = INDEX
  ISUM = 0
  DO  J = 1, N
    IF  ( ICOL1 ( J ) . NE . 0 )  THEN
        IF  ( J . GT . 1 )  ISUM = ISUM + ICOL1 ( J - 1 )
        ISTART = ISUM + IL ( J )
        IF  ( ISTART . EQ . ( ISUM + ICOL1 ( J ) + 1 ) )  GOTO 10
        K = IND1 ( ISTART )
        CALL  DECOMIJ ( K, M, N, I I, JJ )
        IF  ( I I . EQ . I ) THEN
          NEXT ( INDEX ) = ISTART
          INDEX = INDEX + 1
          IL ( J ) = IL ( J ) + 1
        END  IF
    END  IF
10        CONTINUE
  END  DO
  ICOL2 ( I ) = INDEX - INDEXHOLD
END  DO
CALL  LIB$SHOW-TIMER
DO  L = 1, 363205
  IF  ( IND1 ( NEXT ( L ) ) . EQ . 0 ) GOTO 15
  K = IND1 ( NEXT ( L ) )
  CALL  DECOMIJ ( K, M, N, J, I )
  WRITE ( *, 20 )  I, J, A ( NEXT ( L ) )
20      FORMAT ( 2 ( ' ' , I 3 ) , ' ' , D22.15 )
END  DO
15    CONTINUE
END
SUBROUTINE  DECOMIJ ( K, M, N, I, J )
J = ( K - 1 ) / M + 1
I = K - M * ( J - 1 )
RETURN
END
```

Figure 3.5. In-place transpose of sparse matrix stored as paired vectors.

```
# define NULL  0                                          /* End if list indicator */
# define BEGIN  0                                         /* Beginning of list indicator */
# define END-OF-FILE  -999                                /* End of data indicator */
# define SWAP (x, y, t) ((t)) = (x), (x) = (y), (y) = (t))  /* Macro to interchange values */
struct matrix                                             /* Structure template for the linked list */
{                                                         /* representation of sparse matrix */
    short int ind1 ;
    short int ind2 ;
    double aij ;
    long int * ichain ;
} ;
main ( )
{
    struct matrix * mat ;                                /* * mat is pointer to matrix elements */
    long int * start , * oldptr , * iptr ;
    double aij , atrij ;
    int i , j , k , m , n , temp ;
    char * malloc ( ) ;
    printf ("m = ?\n") ;
    scanf ("%d", & m) ;
    printf ("n = ?\n") ;
    scanf ("%d", & n) ;
    k = BEGIN ;
    printf ("i = ?\n") ;
    scanf("%d", & i) ;
    while ( i ! = END-OF-FILE )
    {
        iptr = (int *) malloc (16) ;                     /* Allocate space for matrix element */
        mat = iptr ;
        mat -> ind1 = i ;
        printf ("j = ?\n") ;
        scanf ("%d", & j) ;
        mat -> ind2 = j ;
        printf ("aij = ?\n") ;
        scanf ("%f", & aij) ;
        mat -> aij = aij ;
        if ( k == BEGIN )
            start = iptr ;                               /* Set start equal to first pointer */
        else
        {
            mat = oldptr ;                               /* Otherwise set ichain from previous */
            mat -> ichain = iptr ;                       /* pointer equal to current pointer */
        }
        oldptr = iptr ;                                  /* Keep address of current pointer */
        k ++ ;
        printf ("i = ? \n") ;
        scanf ("%d", & i) ;
    }
    mat = oldptr ;
    mat -> ichain = NULL ;                               /* Indicate end of list */
    printf ("\n") ;
    printf ("Out of %d matrix elements %d are non - zero \n", m * n, k) ;
    printf ("\n") ;
    printf ("The transpose matrix is : \n") ;
    mat = start ;                                        /* Set matrix pointer to first address */
    while ( iptr ! = NULL )
    {
        i = mat -> ind1 ;
        j = mat -> ind2 ;
        atrij = mat -> aij ;
        SWAP (i, j, temp) ;
        printf ("i = %d  j = %d  atrij = %22.15e \n", i, j, atrij) ;
        iptr = mat = mat -> ichain ;                     /* Use ichain to point to next address */
    }
}
```

Figure 3.6. Representation of sparse matrix as linked list.

```
C
C     USE HASH TABLE TO REPRESENT A SPARSE MATRIX. PARAMETER IB IS THE SIZE OF THE TABLE.
C     IT SHOULD BE SET TO 4 * J + 3 , WHERE J IS PRIME.
C
              PROGRAM VMHASH
              REAL *8 AIJ , ATRIJ , SUM
              INTEGER *2  I , J , ITEMP , IND (59)
              REAL *8   A (59)
              DATA IB / 59 /
              OPEN ( 1 , FILE = 'INDATA.DAT ', STATUS = 'OLD ')
              READ ( 1 , 50 ) M , N
50            FORMAT ( 2 ( I 2 , ' ' ))
              KOUNT = 0
1             READ ( 1 , 55 ) I , J , AIJ
55            FORMAT ( 2 ( I 3 ) , D22.15 )
C     -999 SIGNALS THE END OF THE INPUT
              IF ( I . EQ. -999 ) GOTO 25
              KOUNT = KOUNT + 1
              K = M * ( J - 1 ) + I
              IK = MOD ( K , IB ) + 1
C     CONSTRUCT HASH TABLE USING LINEAR PROBING
              IF ( IND ( IK ) . EQ. 0 ) THEN
                 IND ( IK ) = K
                 IC = IK
              ELSE
C     IF BUCKET ALREADY OCCUPIED SEARCH FOR FIRST AVAILABLE BUCKET
                 DO  L = IK + 1 , IB
                    IF ( IND ( L ) . EQ . 0 ) THEN
                       IND ( L ) = K
                       IC = L
                       GOTO 15
                    END IF
                 END DO
C     WRAP TABLE AROUND IF NO AVAILABLE SLOT FOUND BEFORE UPPER LIMIT
                 DO  L = 1 , IK - 1
                    IF ( IND ( L ) . EQ . 0 ) THEN
                       IND ( L ) = K
                       IC = L
                       GOTO 15
                    END IF
                 END DO
C     TERMINATE PREMATURELY IF HASH TABLE FILLS UP
                 TYPE * , 'HASH TABLE IS FULL '
                 STOP
              END IF
15            CONTINUE
              A ( IC ) = AIJ
              GOTO 1
25            TYPE * , 'OUT OF ', M * N , 'ELEMENTS ', KOUNT , 'ARE NON-ZERO '
              TYPE * , '
C     CALCULATE EUCLIDEAN NORM OF THE ROWS
              DO I = 1 , M
```

Figure 3.7. Calculation of Euclidean row norm of sparse matrix using hashing.

```
              SUM = 0.0
              DO J = 1 , N
                K = M * ( J - 1 ) + I
                CALL HASHSRCH ( K , IND ( 1 ) , IC )
                IF ( IC . NE . 0 ) THEN
                  AIJ = A ( IC )
                  SUM = SUM + AIJ ** 2
                END IF
              END DO
              SUM = SQRT ( SUM )
              TYPE *, 'THE EUCLIDEAN NORM OF ROW    ', I , 'OF A IS ', SUM
            END DO
            END
C    SUBROUTINE  FOR CONDUCTING THE HASH SEARCH WITH LINEAR PROBING
            SUBROUTINE HASHSRCH ( K , IX , IC )
            INTEGER * 2 IX ( 1 )
            DATA IB / 59 /
            IC = 0
            IK = MOD ( K , IB ) + 1
C    ELEMENT FOUND ON FIRST TRY
            IF ( IX ( IK ) . EQ . K ) THEN
              IC = IK
              RETURN
            ELSE
C    IF NOT SEARCH FOR ELEMENT OR  0  (ELEMENT NOT IN TABLE )
              DO I = IK + 1 , IB
                IF ( IX ( I ) . EQ . 0 ) THEN
                  IC = 0
                  RETURN
                END IF
                IF ( IX ( I ) . EQ . K ) GOTO 5
              END DO
C    WRAP TABLE AROUND
              DO I = 1 , IK - 1
                IF ( IX ( I ) . EQ . 0 ) THEN
                  IC = 0
                  RETURN
                END IF
                IF ( IX ( I ) . EQ . K ) GOTO 5
              END DO
              IC = 0
              RETURN
            END IF
5           CONTINUE
            IC = I
            RETURN
            END
```

Figure 3.7 (*continued*)

References

George, A. and Liu, J.W. (1981). *Computer Solution of Large Sparse Positive Definite Systems* (Prentice-Hall, Englewood Cliffs, N.J.).

Horowitz, E. and Sahni, S. (1984). *Fundamentals of Data Structures in PASCAL* (Computer Science Press, Rockville, Md.).

Knuth, D.E. (1973). *Sorting and Searching*: Vol. 3 of *The Art of Computer Programming* (Addison-Wesley, Reading, Mass.).

Pooch, U.W. and Nieder, A. (1973). A Survey of Indexing Techniques for Sparse Matrices, *ACM Computing Surveys*, **5**, No. 2, p. 109.

Wirth, N. (1976). *Algorithms + Data Structures = Programs* (Prentice-Hall, Englewood Cliffs, N.J.). In 1985 a version of this book with programming examples in Modulo-2 instead of PASCAL appeared.

CHAPTER 4

Introduction to Overdetermined Systems

4.1. Introduction

After much preliminary discussion we finally begin in this chapter with the subject of overdetermined systems. In this chapter, and the next two, we assume linear overdetermined systems, a restriction that will be relaxed in Chapter 7. Our system is

$$\mathbf{A} \cdot \mathbf{X} = \mathbf{d}, \tag{4.1}$$

where $\mathbf{A}$ is a matrix, called the data matrix or matrix of equations of condition, of size $m \times n$, with $m \geq n$, $\mathbf{X}$ is an n vector of the desired solution, and $\mathbf{d}$ is an m vector of observations or experimental data points.

Before discussing Eq. (4.1) we should first ask how it is that overdetermined systems arise in the first place. They are a rather recent development in the long history of mathematics. The engineers responsible for building the seven wonders of the ancient world used mathematics to aid them in construction, but there is no evidence that they used overdetermined systems. Even today, navigators do not rely on overdetermined systems to establish the position of a ship. Rather, when employing celestial navigation, they use celestial observations to obtain two lines of position whose intersection gives the position of the vessel. Often a third line of position is obtained that, because of measurement errors, does not intersect at the same point but forms a triangle with the previous two lines of position. The vessel is assumed to lie within the triangle, whose size gives some idea of the uncertainty of the position. Perhaps a bit of history of overdetermined systems may be useful.

To solve Eq. (4.1) at the very least m must equal n, which gives a unique solution. If $m > n$ and the equations are consistent, we can select any subset n of the m equations, solve them, and the solution is valid for the remaining equations. The overdetermined system is unnecessary because $(m - n)$ equations are superfluous. But in problems arising from the analysis of experimental data, which are never perfect, the equations are inconsistent; no one vector $\mathbf{X}$ will satisfy all of the equations, but gives rise to residuals

$$\mathbf{r} = \mathbf{A} \cdot \mathbf{X} - \mathbf{d}. \tag{4.2}$$

We want to select an **X** that makes **r**, in some sense, small. Because often an unknown of interest is smaller than the experimental error of the data, only by incorporating many equations of condition into the system, and relying on the statistical principle of beating down measurement error by combining many measurements, can we hope that a reliable determination of the unknown will result. In modern research we use thousands of equations to determine ten or twenty unknowns.

It was not always so. In the ancient and medieval world and up to the seventeenth century in Europe—there is some indication that the Arabs made continuous series of observations—one obtained only enough observations to allow them to find the unknown of interest. In effect, one took $m = n$. The most famous astronomer of antiquity, Claudius Ptolemy (c. 100–c. 178), states this philosophy well in *The Almagest* (1952, p. 296):

"... we first took this way of finding the whereabouts on the ecliptic of Mercury's apogee. For we examined the observations of the greatest elongations in which the morning positions of the star [Mercury; the word "planet" derives from the Greek for "wandering star"] were equal to the evening ones in their angular distance from the mean position of the sun. When these had been found, it followed from what we have shown that the point midway between the two positions included the eccentric's apogee."

There is no hint here of using all of the observations, or even a substantial subset of them, to estimate the eccentric's apogee, only the selected observations needed to determine the quantity. Given the relative crudity of ancient planetary models, some of Ptolemy's predicted planetary positions are in error by an amount equivalent to four times the diameter of the full moon, such a parsimonious observation philosophy was nevertheless sufficient.

The Danish astronomer Tycho Brahe (1546–1601) was the first, at least in Europe, to make an extended series of observations, not merely to estimate a particular quantity of interest but to provide an observational data base for general use. Johannes Kepler (1571–1630), Imperial Mathematician of the Holy Roman Empire (a position that he remarked as having a magnificent title and miserable salary), used Brahe's observations of Mars to deduce his three laws of planetary motion, a feat that would have been impossible with the pre-Brahe observation philosophy.

But even Kepler did not solve overdetermined systems. He used some of the observations to determine certain parameters, which were later checked and perhaps modified by use of the other observations. But as the observations became more accurate—during the seventeenth century the telescope began to replace instruments that relied on the naked eye—planetary models became more complicated. The simple ellipse for planetary motion, deduced by Kepler, no longer sufficed. Planets were found to follow complicated, nonclosing orbits for which an ellipse is only a first approximation. It then became necessary not only to use an extensive series of observations, but to reduce them by use of an overdetermined system.

Apparently the first technique used, the method of averages attributable to the German astronomer Tobias Mayer (1723–1762) was based solely on convenience rather than mathematical theory. We have m equations and n unknowns with $m > n$. We separate the m equations into n groups, sum the equations within each group, and solve the resulting $n \times n$ system. But there are $m!/(m - n)!$ possible ways to form the $n \times n$ system. Some groupings give good results, where "good" means results close to those obtained from the method of least squares, and others bad or even atrocious results. Scarborough (1966, pp. 528–533) presents some details on the method of averages, but it is really of historical interest only. No sane scientist would consider the method of averages for the computer solution of overdetermined systems.

Before proceeding the author should clarify one point. The reader may be surprised at the number of references to astronomers in this historical discussion. Is this an instance of overenthusiasm, perhaps even fanaticism, on the part of the author, who happens to be an astronomer? Not really. Astronomy was the first of the physical sciences to be faced with the problem of analyzing relatively precise data. Even pretelescopic observations of planetary positions are capable of three parts in ten thousand accuracy, although Tycho Brahe was the first to approach such a high standard; both Ptolemy and Copernicus contented themselves with far cruder accuracy. With a telescope the accuracy jumps to several parts in a million. Furthermore, the quantities of interest must often be determined from observations made over a limited portion of the planet's orbit from a limited portion of the Earth's orbit. The resulting equations are ill-conditioned and we must carry many decimals in the calculations. When discussing Brahe's pretelescopic observations of Mars, Kepler found himself computing by hand with numbers of eight and nine decimals. At one point in his *Commentaries on the Motions of Mars*, published in 1609, he bursts out with, "... if you, dear reader, are bored with this tedious method of calculation, pity me who had to repeat it at least seventy times, losing much time...." Today we shove the menial chore of computing onto the computer, but in the precomputer era and given the relative accuracy of the observations and lengthy calculations, it is small wonder that astronomers were prime contributors to practical numerical methods, including the solution of overdetermined systems.

4.2. Mathematical Theory of Overdetermined Systems

Of the methods for solving Eq. (4.1), that depend on something more than obtaining a solution by any seat of the pants technique, three are of importance and derive from the general principle

$$\|\mathbf{r}\|_p = \|\mathbf{A} \cdot \mathbf{X} - \mathbf{d}\|_p = \min, \tag{4.3}$$

or minimize the pth norm of the vector of residuals. If we set $p = 1$ we obtain an L_1 solution, historically called the method of minimum sum. $p = 2$ corresponds to the famous criterion of least squares, and $p = \infty$ gives an L_∞ solution, also known as a Chebyshev or min–max solution. The L_∞ solution was referred to as a minimum approximation to Eq. (4.1) (Whittaker and Robinson, 1967, pp. 258–259). Other choices for p are possible, but seldom used.

There is little doubt that of the three criteria, L_1, L_2, L_∞, the least squares criterion is, for various reasons, by far the most important. But before discussing these reasons we should mention briefly the other two criteria, although the L_1 method will be discussed extensively in Chapter 6.

Both Laplace (1749–1827), in 1799 in his monumental *Celestial Mechanics*, and Legendre (1752–1833), in 1806 in his *New Methods for Determining Orbits of Comets*, proposed an L_∞ solution of Eq. (4.1). Remembering the discussion of vector norms in Chapter 2, we realize that this method minimizes the largest residual of Eq. (4.2). Such a criterion is useful in some applications where we do not want the error of an approximation, as measured by the size of the residuals, to exceed some pre-established limit. An example is the approximation of a mathematical function by a power or other series. For details, see Fike (1968, Chaps. 5 and 9). Both the L_1 and L_2 methods permit isolated large residuals. But for the analysis of experimental data the L_∞ criterion must be rejected because it assigns high weight to large residuals, which have an undue influence on the solution. In modern terminology the L_∞ criterion is unrobust. We shall have more to say later on robustness.

The English statistician F.Y. Edgeworth—for more recent historical figures no dates of birth or death will be given—in 1887 proposed the L_1 criterion as a generalization of Laplace's method of situation (Edgeworth, 1887a). The criterion minimizes the sum of the absolute values of the residuals and is extremely robust, that is, little influenced by large residuals. The L_1 method is so useful for the analysis of experimental data that all of Chapter 6 will be devoted to it.

Despite the utility of the L_1 method, the method of least squares is still the undisputed champion of reduction methods for overdetermined systems. Why? Historically, the most important reason is the tractability of the mathematics. Legendre, who advocated the Chebyshev criterion, nevertheless found that obtaining a solution was so cumbersome that he abandoned the L_∞ method for the method of least squares. Gauss (1777–1855) remarked, in 1809 in his *Theory of the Motion of the Heavenly Bodies Moving about the Sun in Conic Sections* (1963, pp. 269–270):

"In this case [more equations than unknowns] care must be taken to establish the best possible agreement, or to diminish as far as practicable the differences. This idea, however, from its nature, involves something vague. For, although a system of values for the unknown quantities which makes *all* the differences respectively less than another system, is without doubt to be preferred to the latter, still the choice between two systems, one of which presents a better agreement in some observations, the other in others, is left in a measure to our judgment, and innumerable different principles can be proposed by which the former condition is satisfied ... . But of all these principles ours is the most simple; by the others we should be led into the most complicated calculations."

Notice that in Gauss's shrewd comments there is no indication that a least squares solution is, in some sense, "better" than a solution given by another criterion, only simpler.

Let us see where some of the convenience of least squares comes from. In the next chapter, devoted entirely to least squares, we present a matrix derivation of the normal equations, which exhibits clearly some of the properties of the matrix. For the moment we use a simple derivation obtained from calculus. Suppose we have a group of m data points d_i and want to fit a linear model with n parameters to the data:

$$f_i(a_{i1}, a_{i2}, \ldots, a_{in}) = \sum_{j=1}^{n} a_{ij}X_j. \tag{4.4}$$

A residual is

$$r_i = \sum_{j=1}^{m} a_{ij}X_j - d_i, \tag{4.5}$$

and from the principle of least squares we desire

$$\sum_{i=1}^{m} r_i^2 = \min. \tag{4.6}$$

From (4.5) using calculus we obtain

$$\frac{\partial(\sum_i r_i^2)}{\partial X_j} = 2\sum_i \left(\sum_j a_{ij}X_j - d_i\right)a_{ij} = 0. \tag{4.7}$$

Had we adopted any power other than two in the criterion (4.6) the equation corresponding to Eq. (4.7) would not be linear; the expression within parentheses would be raised to powers other than unity. Its solution, particularly in the era of computations with logarithms or even a mechanical calculating machine, would be extremely cumbersome.

A study of Eq. (4.7) also shows that our resulting linear system is of size $n \times n$:

$$\begin{pmatrix} \sum a_{i1}^2 & \sum a_{i1}a_{i2} & \cdots & \sum a_{i1}a_{in} \\ \sum a_{i2}a_{i1} & \sum a_{i2}^2 & \cdots & \sum a_{i2}a_{in} \\ & & & \vdots \\ \sum a_{in}a_{i1} & \sum a_{in}a_{i2} & \cdots & \sum a_{in}^2 \end{pmatrix} \begin{pmatrix} X_1 \\ X_2 \\ \vdots \\ X_n \end{pmatrix} = \begin{pmatrix} \sum y_i a_{i1} \\ \sum y_i a_{i2} \\ \vdots \\ \sum y_i a_{in} \end{pmatrix}. \tag{4.8}$$

(In Eq. (4.8) all sums run from 1 to m.) The compression of a set of $m \times n$ equations to one of size $n \times n$, by sequential accumulation of the equations of condition, was an important factor in the precomputer era and one not to be despised even with a computer. And here, unlike the method of averages, the $n \times n$ system is unique.

Is ease of use the only reason for adopting the least squares criterion? Since the time when Gauss made his remarks in 1809 many attempts have been made, including by Gauss himself, to demonstrate that the least squares criterion is somewhat more fundamental than other criteria. Textbooks on statistics give two important justifications for the principle of least squares: the Gauss–Markov theorem; and applying maximum likelihood estimation to a data set whose observational errors are distributed according to the normal law.

This is not a textbook on statistics, and we shall assume that the reader is familiar with the basic concepts of the discipline. Nevertheless, for the sake of completeness we present some definitions. The normal frequency distribution of a continuous variable X is

$$f(X) = \frac{1}{\sigma\sqrt{2\pi}} e^{-(X-X_0)^2/2\sigma^2}, \tag{4.9}$$

where X_0 is the center of the distribution and σ, the standard deviation, is a measure of the concentration of the distribution: small σ means a narrow distribution, large σ a wide one. Figure 4.1 shows a graph of the normal distribution.

In Eq. (4.9) it is perhaps best to think of σ as a parameter of the distribution. For a group of discrete values, $X_1, X_2, \ldots, X_n$, with a tendency to cluster around a central value we define the mean

$$X_0 = \frac{1}{n} \sum_{j=1}^{n} X_j, \tag{4.10}$$

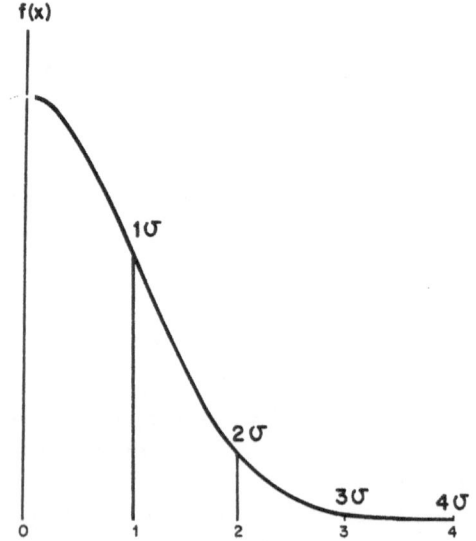

Figure 4.1. The normal distribution.

and the variance

$$\text{var} = \frac{1}{n-1} \sum_{j=1}^{n} (X_j - X_0)^2, \tag{4.11}$$

whose square root is the standard deviation σ, and the mean absolute deviation (MAD)

$$\text{MAD} = \frac{1}{n} \sum_{j=1}^{n} |X_j - X_0|. \tag{4.12}$$

The Gauss–Markov theorem on least squares, proved in books on mathematical statistics, states that within the class of linear, unbiased estimators the least squares estimator has minimum variance. Bias is an intuitive concept that is difficult to define precisely. An unbiased estimator is one that on average gives the results we expect it should. If we flip a coin one hundred times we expect about fifty heads and fifty tails. An estimator that predicted twenty heads and eighty tails would be biased. The mean (4.10) is unbiased, but if we subtract a constant from it, $X_0' = X_0 - 5$, X_0' is a biaseed estimator. Notice that the Gauss–Markov theorem does not restrict the distribution law of the errors of the observed quantities. In particular, it does not assume that the errors follow the normal law.

But it does restrict the class of estimators to linear, unbiased ones. Aside from unusual circumstances, it is hard to imagine why one would desire a biased estimator. At times, however, a nonlinear estimator may be preferable to a linear one. As an example, consider finding the average of a series of numbers. We generally take the mean, a linear estimator, as the average. But just one discordant number in the series can influence the mean unduly. In such an instance the median, a nonlinear estimator little bothered by discordant data, gives a more realistic average.

The Gauss–Markov theorem also assumes that all of the experimental error is concentrated in the vector **d**. **A** is assumed error free. Should this assumption prove unjustified, then the Gauss–Markov theorem is inapplicable. Fortunately, considering **A** error free is usually defensible because it is calculated from a known mathematical model. The model, of course, may be unrealistic, but once it is adopted the elements of **A** are known exactly. They are obtained by calculation from the model and uncorrupted by inevitable experimental error.

Further mathematical justification for the least squares criterion comes from maximum likelihood estimation. This technique for parameter estimation, explained fully in statistics texts, may be described briefly in the following way. We have a sample of m data points—the components of the vector **d** of Eq. (4.1)—to be explained by a mathematical model with n parameters—the vector **X** of Eq. (4.1). In Eq. (4.1) the relationship between **X** and **d** is expressed linearly, but maximum likelihood estimation in its most general formulation merely considers that **d** comes from a probability distribution that is specified by the parameters **X**; represent it by $p(\mathbf{d}, \mathbf{X})$. The probability distribution is general. If the components of $\mathbf{d}$ are independent the general probability distribution may, by the principle of multiplication of independent probabilities, be expressed as

$$p(\mathbf{d}, \mathbf{X}) = p(d_1, \mathbf{X})p(d_2, \mathbf{X}) \ldots p(d_m, \mathbf{X}). \tag{4.13}$$

The likelihood function, $L(\mathbf{X})$, is merely the right-hand side of Eq. (4.13),

$$L(\mathbf{X}) = \prod_{i=1}^{m} p(d_i, \mathbf{X}), \tag{4.14}$$

and we ask: Which values of $\mathbf{X}$ maximize the likelihood function $L(\mathbf{X})$? From elementary calculus this happens when

$$\frac{\partial L}{\partial X_1} = \frac{\partial L}{\partial X_2} = \cdots = \frac{\partial L}{\partial X_n} = 0. \tag{4.15}$$

Let us now restrict the probability distribution to the normal distribution. In particular, let us write, for the probability distribution of the errors of Eq. (4.1),

$$p(d_i, \mathbf{X}) = \frac{1}{\sigma\sqrt{2\pi}} \exp\left[\frac{-(d_i - \sum_{j=1}^{n} a_{ij}X_j)^2}{2\sigma^2}\right] \tag{4.16}$$

The expression within parentheses in the exponent is the residual r_i. We have for our likelihood function

$$L(\mathbf{X}) = \prod_{i=1}^{m} \frac{1}{\sigma\sqrt{2\pi}} e^{-r_i^2/2\sigma^2}$$

$$= \left(\frac{1}{\sigma\sqrt{2\pi}}\right)^m \exp\left(-\sum_{i=1}^{m} \frac{r_i^2}{2\sigma^2}\right). \tag{4.17}$$

An exponential obtains its maximum when the exponent is a minimum. Because we have considered σ a constant we obtain

$$\sum_{i=1}^{m} r_i^2 = \min,$$

the least squares criterion. If, therefore, the observational errors follow the normal law, maximum likelihood estimation coincides with least squares estimation.

But do observational errors follow the normal law? Here opinions run the gamut from considering the normal law as a fundamental law of nature to which observational data, with few exceptions, conform to regarding the normal law as a superfluous hypothesis to be dispensed with as much as possible. Representative of the former point of view is Scarborough (1966, p. 491):

The truth is that, for the kinds of errors considered in this book (errors of measurement and observation), the Normal Law is *proved by experience* [italics in original]. Several substitutes for this law have been proposed, but none fits the facts as well as it does.

Regarding the latter point of view we read in the recent book by Press, Flannery, Teukolsky, and Vetterling (1986, p. 501):

For a hundred years or so, mathematical statisticians have been in love with the fact that the probability distribution of the sum of a very large number of very small random deviations always converges to a normal distribution.... This infatuation tended to focus interest away from the fact that, for real data, the normal distribution is often rather poorly realized, if it is realized at all.

That the sum of a large number of small random deviations converges to the normal distribution is called, in mathematical statistics, the central limit theorem.

It seems to the author that both of these extreme viewpoints are unwarranted. Certainly, we need not search far to find observational distributions that are poorly approximated by the normal distribution. Figure 4.2, given in one of the author's papers (Branham, 1986), shows the distribution of the residuals from 21,365 observations of five minor planets. The distribution

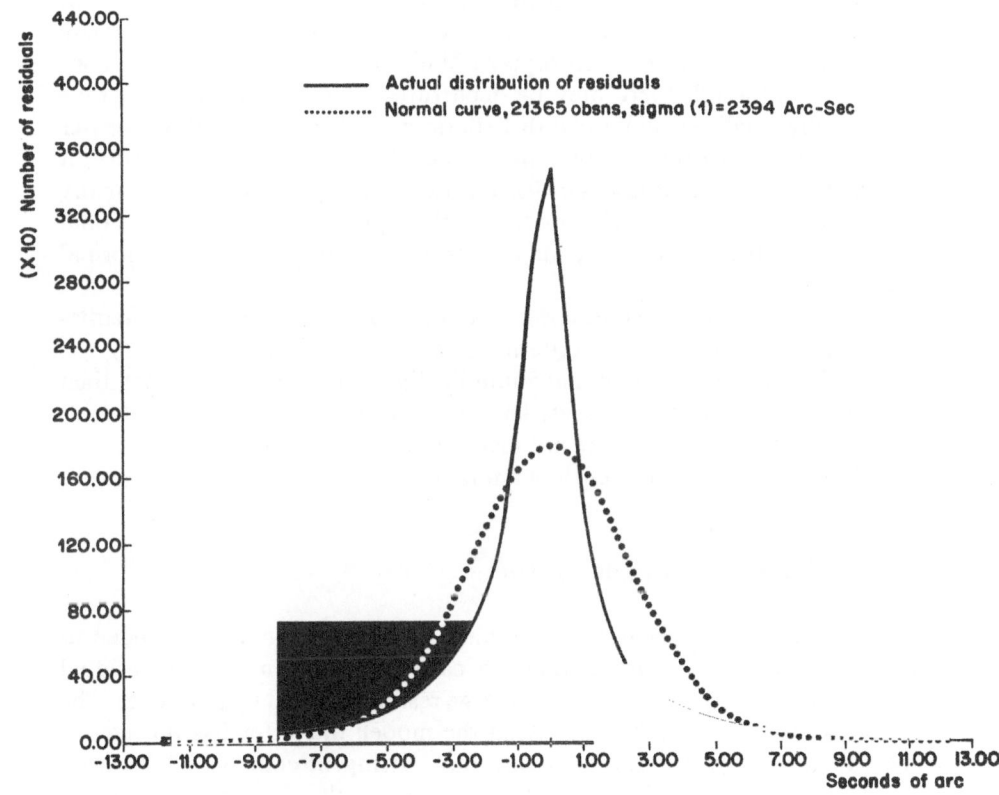

Figure 4.2. Distribution of 21,365 residuals from five minor planets.

is more peaked (leptokurtic in the language of statistics) and heavy tailed than the normal distribution. The distribution is also skewed; there are more positive than negative residuals.

On the other hand, those who assert that the normal law "...is often rather poorly realized, if it is realized at all" should look at Section 2.20 of Smart's book (1958). Smart takes 470 observations of the stars Sirius and Altair and compares the distribution of error in one of the astronomical coordinates with the normal law. He concludes: "The agreement between the implications of the normal law and the observational errors ... is remarkably close on the whole".

What, then, is the status of the normal law? The confusion of thought on the matter once led the French physicist Gabriel Lippmann to remark, perhaps humorously, perhaps sarcastically, to Poincaré (Whittaker and Robinson, 1967, p. 179): "Everybody believes in the exponential law of errors [the normal law]: the experimenters because they think it can be proved by mathematics; and the mathematicians, because they believe it has been established by observation." Whittaker and Robinson themselves provide a balanced statement of the matter on the same page: "Theory asserts, and observation confirms the assertion, that the normal law is to be expected in a very great number of frequency distributions, but not in all." A modern statistician, Stigler (1977), finds evidence to support Whittaker and Robinson. He concludes that, although there may be departures from the normal law, they are generally small—usually real distributions tend to have slightly heavier tails—and show no remarkable abnoralities. On weighing the evidence, it appears as if the normal law is at least a useful first approximation in many or even most experimental situations, but there nevertheless exist occasional observational distributions that depart, perhaps seriously, from the normal law.

To summarize, the least squares criterion is supported by more mathematical theory than the other criteria. If the assumptions of the theory, the Gauss–Markov theorem and maximum likelihood estimation, are met, then a least squares solution of an overdetermined system gives an optimal solution. But if the assumptions are not met, there is no reason to suppose that a least squares solution is better than another given by a different criterion.

4.3. Modeling Errors and Outliers

In Section 4.2, when discussing the Gauss–Markov theorem, we alluded to modeling errors. We must distinguish carefully between a large residual caused by an inadequate model and a large residual caused by poor data. The former can be remedied by improving the model; the latter has no remedy unless the experiment is repeated with more stringent controls on the accidental errors, if this is possible. With regard to modeling errors, we must also distinguish between curve fitting of an arbitrary model with as many adjustable parameters as we need, and fitting a known model, known from physical considerations, for example, to the data.

With fitting adjustable parameters it is always possible to have a perfect fit: all residuals are zero. How? If we have m data points we pass a polynomial of degree $m - 1$ through them. The polynomial, of course, will oscillate wildly between data points and will be totally useless for interpolation between them or limited extrapolation beyond the ends of the interval. The correct procedure for curve fitting passes a relatively low-order mathematical function, usually a polynomial, through the data points. If statistical tests indicate remaining, unexplained variance, then the order of the polynomial may be increased. As soon as statistical tests show that, given the error of the data, no significant, unexplained variance remains, the fitting is over. The resulting polynomial will be useful for interpolating between data points and for limited extrapolation.

If we know that the data are to be explained by a given physical model, then curve fitting with arbitrary parameters may not be used. Suppose, for example, that we wish to determine the force constant of an elastic material. As long as the elastic limit is not exceeded, Hooke's law assures us that the equation relating deformation to the magnitude of the applied force is linear. Here we may not fit an arbitrary number of parameters even if statistical tests show a valid decrease in variance if we add quadratic, cubic, or other terms. Because we know that the law is linear any such decrease in variance would be attributable to systematic errors in the data. Or perhaps we have selected the wrong model; the material is not elastic and a linear model is incorrect. But once a model is chosen, the number of parameters is specified and we have no right to arbitrarily add or subtract parameters. In scientific terms, the curve fitting of the previous paragraph corresponds to finding an empirical formula, and the considerations of this paragraph to determining the constants of a physical law.

Even if our model is adequate we may, nevertheless, encounter large residuals. Poor data, of course, give rise to large residuals, but regardless of the quality of the data some residuals are so large that they cannot be explained by chance fluctuations in the data. If we are using a criterion, such as that of least squares, that is sensitive to large residuals, a few really large residuals can transform a potentially useful solution to nonsense. Suppose that we have data that follow closely the normal law. A three standard deviation (σ) residual will be encountered once in a thousand times, a five σ residual once in three million, and a ten σ residual only once in 10^{23}. Such large residuals, called outliers, almost always corresponds to blunders—perhaps the experimenter bumped the apparatus or recorded a wrong number—and must be eliminated. Occasionally, a large residual represents a physical effect not contemplated in the physical model. During the nineteenth century observations of the position of the planet Mercury showed large residuals that could not be explained. Only with the development of general relativity did it become evident that Newtonian mechanics is inadequate to explain the planet's motion. But, in general, large residuals are outliers and unless they are eliminated the method of least squares, considering them genuine chance fluctuations, will contort the entire solution to bring them into line. The L_1 method, on the other hand, is little bothered by outliers.

But exactly how can we determine what is an outlier? No one would have qualms about eliminating a ten σ residual, but what about a four σ residual? Should that be considered a genuine data point—on a chance basis it would be encountered three times in ten thousand—or an outlier? Outlier rejection is really a form of robust estimation, of which we shall say more in the next chapter. The subject has a long and interesting history. Although a century-old Edgeworth's paper (1887b) on discordant observations still repays careful study.

From the time least squares was first used observations considered discordant were eliminated from further consideration on a more or less subjective basis. The astronomer Benjamin Pierce, professor of mathematics at Harvard University, appears to have been the first to propose a criterion for outlier rejection based on probability (1852).

Whether it is possible from probability alone to determine discordant observations was the subject of intense and sometimes heated debate. No less an authority than the Astronomer Royal, Sir George Airy, felt that the task was hopeless, that the only valid criterion for outlier rejection comes from external evidence, perhaps from an examination of observers' logbooks (1856). This extreme viewpoint was accepted by few and is obviously inapplicable when the data base is large and has been gathered automatically. But Airy's ideas, nevertheless, emphasize that we should be careful with outlier rejection. As we mentioned previously a large residual may indicate some physical effect not contemplated in the model.

As a compromise position some authorities considered that probability alone offers no justification for the outright rejection of an observation, only that the large residual indicates that the observation is improbable and should be assigned small weight. Edgeworth's (1887b) paper discusses the merits of this and other points of view.

But the general consensus of opinion was, and continues to be, that outliers are blunders and must be eliminated. Pierce (1852) undoubtedly anticipated objections to his criterion when he stated, in words that seem as appropriate today as when first written:

"In almost every true series of observations some are found which differ so much from the others as to indicate some abnormal source of error not contemplated in the theoretical discussions, and the introduction of which into the investigations can only serve, in the present state of science, to preplex and mislead the inquirer."

Edgeworth (1887c) also felt that when we use the least squares criterion—do not forget that he invented the L_1 method—it is best to discard large residuals. He expressed his thoughts in delightful language:

"The Method of Least Squares is seen to be our best course when we have thrown overboard a certain portion of our data—a sort of sacrifice which has often to be made by those who sail upon the stormy seas of probability."

Granted that we desire to eliminate discordant observations, how do we do it? Of the various criteria proposed, that of Pierce is as good as any. Pierce's arguments are subtle—in fact, Edgeworth could not follow the reasoning that allowed Pierce to derive a definite cutoff for a discordant observation—and depend upon the observational errors following the normal law. It is best to regard Pierce's criterion as a convenient rule of thumb rather than a profound consequence of probability theory.

We will present the criterion in a cookbook fashion. Pierce's original article should be consulted for details. We have m data points and wish to eliminate n of them keep $n' : m = n + n'$. Our mathematical model includes μ unknowns, and K is the threshold for rejection. Our algorithm is:

1. For the first approximation take $K = 3\sigma$. The exact value is unimportant, but a three σ filter is commonly used for outlier rejection.
2. Calculate

$$T^n = \frac{n^n n'^{n'}}{m^m}.$$

To avoid problems with overflow we should use the relation

$$n \log T = n \log n + n' \log n' - m \log m$$

and compute T from the antilogarithm.

3. Calculate

$$R = 2e^{(K^2-1)/2}\{1 - P(X = K)\},$$

where $P(X)$ is the probability integral

$$P(X) = \frac{1}{\sqrt{2\pi}} \int_{-\infty}^{X} e^{-t^2/2} \, dt.$$

4. Calculate

$$\lambda^2 = \left(\frac{T}{R}\right)^{2n/(m-n)}.$$

5. Calculate

$$K^2 - 1 = \left(\frac{m - \mu - n}{n}\right)(1 - \lambda^2).$$

6. Use the new value of K and go to step 2. Iterate until convergence.

To have some feel for the cutoffs given by Pierce's criterion these are shown in Table 4.1, which takes a model with two unknowns ($\mu = 2$) and assumes we wish to eliminate one discordant observation ($n = 1$).

Table 4.1 indicates that as the number of observations increases so does the threshold for rejection. This is reasonable as, according to probability theory, large residuals are more likely in large data sets than in small ones. Step 5 of the algorithm shows that as μ increases, K decreases. This is, again, reasonable as more unknowns in the model explain more of the variance, making a large residual less likely. The table also shows that, to the extent that

Table 4.1. Pierce's criterion for $\mu = 2$ and $n = 1$.

m	50	10^2	10^3	10^4	10^5	10^6
K	2.58	2.84	3.59	4.19	4.64	5.08

assumptions in Pierce's criterion, particularly that of normality, are valid, the commonly used three σ cutoff is too restrictive for large data sets. When we analyze data containing thousands or even tens of thousands of observations, a four or five σ filter is more realistic.

4.4. Solution of Linear Systems

Whatever the criterion chosen, L_1, least squares, L_∞, or possibly some other, at some point in the solution of a linear, overdetermined system we need to solve an $n \times n$ linear system, possibly many times. When solving the normal equations arising from the method of least squares we generally use, because of certain favorable features, Cholesky decomposition. But to appreciate the advantages of Cholesky decomposition, discussed in the next chapter, we should first study the method most often used to solve general linear systems, Gaussian elimination, which is also useful in connection with the L_1 method.

Presumably most readers are familiar with Cramer's rule and realize that it is inefficient compared with Gaussian eliminations. But perhaps the point bears one further repetition. Gaussian elimination requires about $n^3/3$ multiplications and divisions to solve an $n \times n$ system. Cramer's rule, expanding the determinants directly by cofactors, requires $n \cdot n!$ multiplications and divisions. For large n, the comparison between $n^3/3$ and $n \cdot n!$ is startling. Should $n = 10$, Gaussian elimination solves our system for the trifling cost of 333 arithmetic operations; profligate Cramer's rule uses up more than 36 million operations. For large n, Cramer's rule is impractical even if we use a more sensible way to calculate the determinants, something other than direct expansion by cofactors.

Gaussian elimination is systematic reduction of a linear system to upper triangular form and solving for the unknowns by back substitution. Every schoolboy quickly solves the equations

$$\begin{aligned} 3X - 4Y + \ Z &= 1, \\ 2X + 5Y \quad\ \ \ &= 1, \\ 8X + 7Y + 9Z &= 1, \end{aligned}$$

by dividing the first equation by three and subtracting the first from the second and third, eliminating X from the second and third equations. From the resulting third equation, containing only Z, he calculates, to three decimals,

$Z = -0.309$. From this value and the second equation, containing only Y and Z, he calculates $Y = 0.017$. And from the first equation he obtains $X = 0.459$.

This familiar procedure in matrix notation consists of taking a linear system

$$\mathbf{C} \cdot \mathbf{X} = \mathbf{b} \tag{4.18}$$

(to avoid confusion with our previous notation for an overdetermined sytem we express the matrix of coefficients by $\mathbf{C}$ rather than $\mathbf{A}$ and the vector of the right-hand side by $\mathbf{b}$ rather than $\mathbf{d}$) and reducing it to

$$\mathbf{U} \cdot \mathbf{X} = \mathbf{c}, \tag{4.19}$$

where $\mathbf{U}$ is upper triangular. To see how $\mathbf{U}$ is formed we ask: What matrix $\mathbf{M}_1$ will eliminate the unknowns X_2 and X_3 in the first column of the system

$$\begin{pmatrix} 3 & -4 & 1 \\ 2 & 5 & 0 \\ 8 & 7 & 9 \end{pmatrix} \begin{pmatrix} X_1 \\ X_2 \\ X_3 \end{pmatrix} = \begin{pmatrix} 1 \\ 1 \\ 1 \end{pmatrix} \tag{4.20}$$

(the same system that the schoolboy solved)? The matrix $\mathbf{M}_1$ is a modified unit matrix, modified by insertion of elements in the kth column to eliminate the subdiagonal elements in the kth column of the product of $\mathbf{M}_1$ with $\mathbf{C}$ and the successively transformed $\mathbf{C}$.

For Eq. (4.20) the matrix $\mathbf{M}_1$ is

$$\mathbf{M}_1 = \begin{pmatrix} 1 & 0 & 0 \\ -\frac{2}{3} & 1 & 0 \\ -\frac{8}{3} & 0 & 1 \end{pmatrix}$$

and results in the system

$$\begin{pmatrix} 3 & -4 & 1 \\ 0 & 7\frac{2}{3} & -\frac{2}{3} \\ 0 & 17\frac{2}{3} & 6\frac{1}{3} \end{pmatrix} \begin{pmatrix} X_1 \\ X_2 \\ X_3 \end{pmatrix} = \begin{pmatrix} 1 \\ \frac{1}{3} \\ -1\frac{2}{3} \end{pmatrix}.$$

The matrix $\mathbf{M}_2$, to eliminate X_2 from the new third equation, is

$$\mathbf{M}_2 = \begin{pmatrix} 1 & 0 & 0 \\ 0 & 1 & 0 \\ 0 & -2.304 & 1 \end{pmatrix}$$

and results in the upper triangular system

$$\begin{pmatrix} 3 & -4 & 1 \\ 0 & 7\frac{2}{3} & -\frac{2}{3} \\ 0 & 0 & 7.870 \end{pmatrix} \begin{pmatrix} X_1 \\ X_2 \\ X_3 \end{pmatrix} = \begin{pmatrix} 1 \\ \frac{1}{3} \\ -2.435 \end{pmatrix}.$$

The solution follows by back substitution. The matrix $\mathbf{U}$ is $\mathbf{M}_2 \cdot \mathbf{M}_1 \cdot \mathbf{C}$ and the matrix $\mathbf{c} = \mathbf{M}_2 \cdot \mathbf{M}_1 \cdot \mathbf{b}$.

Gaussian elimination needs $n - 1$ transformations to reduce an $n \times n$ system to upper triangular form. The matrices M_i are all of unit condition number in the L_2 norm and do not change the condition number of the linear system. But the triangular decomposition will be derailed should a zero develop on the diagonal of C and its transformations. To calculate the elements of M_i we divide by successive diagonal elements, called pivots, of C and its transformations. A zero there causes the algorithm to blow up.

If our system were, for example,

$$\begin{pmatrix} 0 & -4 & 1 \\ 2 & 5 & 0 \\ 8 & 7 & 9 \end{pmatrix} \begin{pmatrix} X_1 \\ X_2 \\ X_3 \end{pmatrix} = \begin{pmatrix} 1 \\ 1 \\ 1 \end{pmatrix}, \tag{4.21}$$

the calculation of M_1 comes immediately to grief upon dividing the two by the zero even though the matrix is not ill-conditioned, $COND(C) = 44$ in the L_1 norm, and has the solution $X_1 = 1.804$, $X_2 = -0.522$, $X_3 = -1.087$.

Not only will a pivot of zero cause immediate disaster, a small pivot, where small is defined relative to the precision of the arithmetic being used, will result in a spectacular erosion of significant digits, not because of the statistical accumulation of round-off error discussed in Chapter 1, but because the division by a small pivot shifts significant bits of the mantissa of our floating-point number into the bit bucket. We are victims, once again, of the computer's finite word length.

The solution to the problem of zero and small pivots consists of merely interchanging rows, in a partial pivoting strategy, or both rows and columns, in a full pivoting strategy, so that large elements go into diagonal positions. Dividing by large pivots avoids the accuracy problems mentioned in the previous paragraph. In matrix notation, pivoting amounts to multiplying the system by a permutation matrix P_i, a matrix whose columns are a permutation of the unit matrix. If, for example, we multiply Eq. (4.21) by the matrix

$$\begin{pmatrix} 0 & 0 & 1 \\ 0 & 1 & 0 \\ 1 & 0 & 0 \end{pmatrix}$$

the third and first rows are interchanged, the desirable eight moves into the diagonal of the system, and M_1 is calculated without difficulty.

If we take pivoting into consideration, the upper triangular matrix of an $n \times n$ system comes from the series of products

$$U = M_{n-1} \cdot P_{n-1} \cdot M_{n-2} \cdot P_{n-2} \cdots M_1 \cdot P_1 \cdot C. \tag{4.22}$$

(The notation P_i refers to the ith application of a permutation matrix, *not* which columns of the identity matrix are permuted; that depends on the application.) Like the M_i's, the P_i's have a unit condition number and do not affect the condition number of the linear system. Their effect, on the contrary, is to stablize Gaussian elimination. Without the P_i's Gaussian elimination is, in general, an unstable algorithm for calculating the solution. With them it is the best algorithm for obtaining the solution of a general linear system.

For a more complete discussion of Gaussian elimination, including AL-GOL, FORTRAN, and PL/I programs, the reader is referred to the classic work by Forsythe and Moler (1967).

References

Airy, G. (1856). Remarks on Pierce's Criterion, *Astron. J.*, **4**, p. 137.

Branham, R.L., Jr (1986). Is Robust Estimation Useful for Astronomical Data Reduction?, *Quarterly J. Royal Astron. Soc.*, **27**, p. 182.

Edgeworth, F.Y. (1887a). A New Method of Reducing Observations Relating to Several Quantities, *Phil. Mag.*, **23**, p. 222.

Edgeworth, F.Y. (1887b). On Discordant Observations, *Phil. Mag.*, **24**, p. 364.

Edgeworth, F.Y. (1887c). The Choice of Means, *Phil. Mag.*, **24**, p. 268.

Fike, C.L. (1968). *Computer Evaluation of Mathematical Functions* (Prentice-Hall, Englewood Cliffs, N.J.).

Forsythe, G. and Moler, C.B. (1967). *Computer Solution of Linear Algebraic Systems* (Prentice-Hall, Englewood Cliffs, N.J.).

Gauss, K.F. (1963). *Theory of the Motion of the Heavenly Bodies Moving about the Sun in Conic Sections* (Dover, New York).

Pierce, B. (1852). Criterion for the Rejection of Doubtful Observations, *Astron. J.*, **2**, p. 161.

Press, W.H., Flannery, B.P., Teukolsky, S.A., and Vetterling, W.T. (1986). *Numerical Recipes: The Art of Scientific Computing* (Cambridge University Press, Cambridge).

Ptolemy, C. (1952). *The Almagest.* Vol. 16: *Great Books of the Western World* (Encyclopedia Britannica, Chicago).

Scarborough, J.B. (1966). *Numerical Mathematical Analysis*, 6th ed. (The John Hopkins Press, Baltimore).

Smart, W.M. (1958). *Combination of Observations* (Cambridge University Press, Cambridge).

Stigler, S.M. (1977). Do Robust Estimators Work with Real Data?, *Annal. Stat.*, **5**, p. 1055.

Whittaker, E. and Robinson, G. (1967). *The Calculus of Observations* (Dover, New York).

CHAPTER 5

Linear Least Squares

5.1. The Normal Equations

In the previous chapter we talked, in general, about the method of least squares and, in particular, the mathematical justification for selecting it over other criteria. In this chapter we consider practical methods for analyzing a least squares problem. Equations (4.4)–(4.8) present a brief derivation, by calculus, of the normal equations, still the most popular—but by no means only—way of solving least squares problems; in Section 5.4 we shall see that orthogonal transformations allow us to obtain a least squares solution without forming normal equations, a procedure that offers certain advantages but also suffers from some drawbacks, something that proponents of orthogonal transformations frequently overlook.

A derivation of the normal equations by matrix techniques, which we now give, has the advantage over the derivation by calculus of showing more clearly the structure of the normal equations. Rewriting Eq. (4.2) as

$$\mathbf{A} \cdot \mathbf{X} = \mathbf{d} + \mathbf{r} \qquad (5.1)$$

and taking derivatives, remembering that $\mathbf{A}$ and $\mathbf{d}$ are constant, we have

$$\mathbf{A} \cdot d\mathbf{X} = d\mathbf{r}. \qquad (5.2)$$

From the least squares criterion

$$\sum_{i=1}^{m} r_i^2 = \min$$

we derive

$$2 \sum r_i \, dr_i = 0 \quad \Rightarrow \quad d\mathbf{r}^{\mathrm{T}} \cdot \mathbf{r} = 0. \qquad (5.3)$$

Combining Eqns. (5.1)–(5.3) we obtain

$$(\mathbf{A} \cdot d\mathbf{X})^{\mathrm{T}} \cdot \mathbf{r} = d\mathbf{X}^{\mathrm{T}} \cdot \mathbf{A}^{\mathrm{T}} \cdot (\mathbf{A} \cdot \mathbf{X} - \mathbf{d}) = 0. \qquad (5.4)$$

Because matrix multiplication is associative

$$d\mathbf{X}^{\mathrm{T}} \cdot (\mathbf{A}^{\mathrm{T}} \cdot \mathbf{A} \cdot \mathbf{X} - \mathbf{A}^{\mathrm{T}} \cdot \mathbf{d}) = 0. \qquad (5.5)$$

X is an arbitrary vector, and the only way for Eq. (5.5) to be true in general is for the parenthesized expression to be zero. Or,

$$\mathbf{A}^T \cdot \mathbf{A} \cdot \mathbf{X} = \mathbf{A}^T \cdot \mathbf{d}. \tag{5.6}$$

Equation (5.6) is the matrix form of the normal equations. **A** is of size $m \times n$, with $m > n$, and $\mathbf{A}^T$ is of size $n \times m$. Therefore, $\mathbf{A}^T \cdot \mathbf{A}$ is of size $n \times n$, and $\mathbf{A}^T \cdot \mathbf{d}$ is a vector of length n. We compress a system of size $m \times n$ to one of size $n \times n$, a considerable saving of space if $m \gg n$. Let

$$\mathbf{B} = \mathbf{A}^T \cdot \mathbf{A}. \tag{5.7}$$

Then,

$$\mathbf{B}^T = (\mathbf{A}^T \cdot \mathbf{A})^T = \mathbf{A}^T \cdot \mathbf{A}^{TT} = \mathbf{A}^T \cdot \mathbf{A} = \mathbf{B}. \tag{5.8}$$

The matrix **B** is symmetric. Not only do we compress our system to one of size $n \times n$ we can, by taking advantage of the symmetry, compress it further to size $n(n + 1)/2$.

The matrix **B** is positive definite or, at worst, nonnegative definite

$$\mathbf{X}^T \cdot \mathbf{B} \cdot \mathbf{X} = \mathbf{X}^T \cdot \mathbf{A}^T \cdot \mathbf{A} \cdot \mathbf{X} = (\mathbf{A} \cdot \mathbf{X})^T \cdot (\mathbf{A} \cdot \mathbf{X}) \geq 0. \tag{5.9}$$

If λ is an eigenvalue of **B**, then

$$\mathbf{B} \cdot \mathbf{X} = \lambda \mathbf{X}. \tag{5.10}$$

Multiplying both the left- and right-hand sides of Eq. (5.10) by $\mathbf{X}^T$ we find

$$\mathbf{X}^T \cdot \mathbf{B} \cdot \mathbf{X} = \mathbf{X}^T \cdot \lambda \mathbf{X} = \lambda \mathbf{X}^T \cdot \mathbf{X} \geq 0. \tag{5.11}$$

Combining Eqns. (5.9) and (5.11) we see that both the left- and right-hand sides of Eq. (5.11) are nonnegative. The eigenvalues, therefore, are all nonnegative. They will only be zero if the vector **X** is null or the matrix **B** is singular. This will in general not be so, and we may say, remembering the definition of a positive definite matrix from Chapter 2, that **B** is positive definite.

Aside from certain special cases, such as **A** being orthogonal, the condition number of **B** is higher than that of **A**, the matrix of the equations of condition. (It is unfortunate that the term "condition" is used in two different senses in the same sentence, but the usage is standard in both instances. The equations of condition impose conditions, in Eq. (4.2) linear conditions, among the unknowns. The condition number measures the linear dependence of the columns of **A**.) A general matrix with real elements, such as **A**, which admits of a decomposition called the singular value decomposition (SVD), is discussed extensively in Chapter 8:

$$\mathbf{A} = \mathbf{U} \cdot \mathbf{S} \cdot \mathbf{V}^T, \tag{5.12}$$

where **U** is an $m \times m$ orthogonal matrix, **V** is an $n \times n$ orthogonal matrix, and **S** is an $m \times n$ matrix with nonzero elements in the diagonal of the upper $n \times n$ submatrix and zeros elsewhere. In other words, the upper $n \times n$ submatrix is a diagonal matrix and the $(m - n) \times n$ rectangular submatrix is a null matrix. The elements $s_1, s_2, \ldots, s_n$ of **S** are called singular values, and the ratio of the largest to the smallest—s_{max}/s_{min} as shown in Chapter 8—is the condition number of **A**. From Eq. (5.7)

$$\mathbf{B} = \mathbf{A}^T \cdot \mathbf{A} = (\mathbf{U} \cdot \mathbf{S} \cdot \mathbf{V}^T)^T \cdot \mathbf{U} \cdot \mathbf{S} \cdot \mathbf{V}^T$$
$$= \mathbf{V} \cdot \mathbf{S} \cdot \mathbf{U}^T \cdot \mathbf{U} \cdot \mathbf{S} \cdot \mathbf{V}^T = \mathbf{V} \cdot \mathbf{S}^2 \cdot \mathbf{V}^T. \tag{5.13}$$

Equation (5.13) is an eigenvalue–eigenvector decomposition of **B** with $\mathbf{S}^2$ containing the eigenvalues and **V** the eigenvectors. In Chapter 2 we demonstrated that the condition number in the L_2 norm of a matrix is the ratio of the maximum to the minimum eigenvalue of the product of the transpose of the matrix with the matrix. From Eq. (5.13) this condition number is s_{max}^2/s_{min}^2. The condition number of the normal equations is the square of that of the equations of condition.

An interesting property of a least squares solution can be inferred from Eq. (5.6) or, more directly, from Eq. (5.4) if we remember that $d\mathbf{X}^T$ is an arbitrary vector

$$\mathbf{A}^T \cdot (\mathbf{A} \cdot \mathbf{X} - \mathbf{d}) = \mathbf{A}^T \cdot \mathbf{r} = 0. \tag{5.14}$$

The vector **r** of the residuals is orthogonal to the column space of **A**.

From Eq. (5.6) we deduce that the solution to the normal equations is given formally as

$$\mathbf{X} = (\mathbf{A}^T \cdot \mathbf{A})^{-1} \cdot \mathbf{A}^T \cdot \mathbf{d}. \tag{5.15}$$

The matrix $(\mathbf{A}^T \cdot \mathbf{A})^{-1} \cdot \mathbf{A}^T$, called the Moore–Penrose pseudoinverse of **A**, coincides with the ordinary inverse $\mathbf{A}^{-1}$ if $m = n$. It is a useful entity when we derive error estimates for the components of the solution vector **X**. If we denote the pseudoinverse by $\mathbf{A}^+$ we can easily derive the conditions, also satisfied by the ordinary inverse:

$$\mathbf{A} \cdot \mathbf{A}^+ \cdot \mathbf{A} = \mathbf{A};$$
$$\mathbf{A}^+ \cdot \mathbf{A} \cdot \mathbf{A}^+ = \mathbf{A}^+;$$
$$(\mathbf{A} \cdot \mathbf{A}^+)^T = \mathbf{A} \cdot \mathbf{A}^+; \tag{5.16}$$
$$(\mathbf{A}^+ \cdot \mathbf{A})^T = \mathbf{A}^+ \cdot \mathbf{A}.$$

For a further discussion of the pseudoinverse—an example of a rectangular matrix having an inverse, mentioned in Chapter 2—see Chapter 7 of Lawson and Hanson's book (1974).

Although matrix notation provides a convenient way to derive the normal equation and exhibit some of their properties, we should not be fooled into thinking that Eq. (5.6) is a practical computational scheme. To form normal equations by explicitly multiplying an $m \times n$ matrix by its transpose would waste valuable computer memory. In practice, we form equations by the sequential accumulation of the rows of **A**. If we use a two-dimensional array for the matrix **B**—not the most efficient representation because it takes no advantage of **B**'s symmetry—FORTRAN coding to form **B** and $\mathbf{A}^T \cdot \mathbf{d}$ from **A** would look something like:

```
DO I = 1, N
  DO J = I, N
    B(I, J) = B(I, J) + CONDEQ (I) * CONDEQ (J)
  END DO
  Y(I) = Y(I) + CONDEQ (I) * RHS
END DO
```

CONDEQ is an array of size n to hold one of the rows of **A** and RHS is a variable with the corresponding component of **d**. Generally, **A** and **d** will be contained in a file on tape or disk. We keep on reading the rows of **A** and component of **d** and accumulating **B** and $\mathbf{Y} = \mathbf{A}^T \cdot \mathbf{d}$ until the file is exhausted.

5.2. Solution of the Normal Equations

Because the matrix **B** is symmetric and positive definite, the normal equations are more efficiently solved by a special technique, Cholesky decomposition— sometimes called the square root or Banachiewicz method, especially in the Russian literature (Faddeeva, 1959, Chap. 2, Sec. 10)—than by Gaussian elimination. In fact, Cholesky decomposition can be considered Gaussian elimination adopted to the special conditions of a symmetric, positive definite matrix.

Forsythe and Moler (1967, pp. 27–28) prove that a general square $n \times n$ matrix **C** may be expressed as

$$\mathbf{C} = \mathbf{L} \cdot \mathbf{U}, \tag{5.17}$$

where **L** is lower triangular with unit elements on the main diagonal and **U** is upper triangular. Equation (5.17) is referred to as the *LU* theorem. Let **U** be expressed as $\mathbf{U} = \mathbf{D} \cdot \mathbf{U}'$, where **D** is a diagonal matrix. If **C** is symmetric, $\mathbf{C}^T = \mathbf{C}$, we obtain from Eq. (5.17) upon using **U**',

$$\mathbf{C}^T = \mathbf{U}'^T \cdot \mathbf{D} \cdot \mathbf{L}^T = \mathbf{C} = \mathbf{L} \cdot \mathbf{D} \cdot \mathbf{U}', \tag{5.18}$$

from which we infer that $\mathbf{U}' = \mathbf{L}^T$. If, furthermore, **C** is positive definite all of the elements of **D** are positive. Why? A theorem of matrix algebra states that if a matrix is positive definite its determinant and the determinant of every submatrix comprising it are positive. If **U**' of Eq. (5.18) is selected so that, like **L**, it has unit elements on its main diagonal then

$$\Delta(\mathbf{C}) = \Delta(\mathbf{L}) \cdot \Delta(\mathbf{D}) \cdot \Delta(\mathbf{L}^T)$$

$$= |\cdot d_{11} \cdot d_{22} \cdots d_{nn} \cdot|, \tag{5.19}$$

where Δ indicates the determinant. Because of the positivity of $\Delta(\mathbf{C})$ and all of its submatrices, all of the d_{ii}'s are positive. If we let $\mathbf{B}$ be $\mathbf{C}$ and set $\mathbf{S}^T = \mathbf{L} \cdot \mathbf{D}^{1/2}$ we obtain, from Eq. (5.17),

$$\mathbf{B} = \mathbf{L} \cdot \mathbf{D} \cdot \mathbf{L}^T = \mathbf{S}^T \cdot \mathbf{S}. \tag{5.20}$$

Equation (5.20) is the Cholesky decomposition of $\mathbf{B}$. In words, if $\mathbf{B}$ is symmetric and positive definite it may be expressed as the product of an upper triangular matrix multiplied by its transpose.

To calculate the elements of $\mathbf{S}$ we start from

$$\begin{pmatrix} B_{11} & B_{12} & \cdots & B_{1n} \\ B_{21} & B_{22} & \cdots & B_{2n} \\ \vdots & & & \vdots \\ B_{n1} & B_{n2} & \cdots & B_{nn} \end{pmatrix} = \begin{pmatrix} S_{11} & 0 & \cdots & 0 \\ S_{12} & S_{22} & & 0 \\ \vdots & \vdots & & \vdots \\ S_{1n} & S_{2n} & \cdots & S_{nn} \end{pmatrix} \begin{pmatrix} S_{11} & S_{12} & \cdots & S_{1n} \\ 0 & S_{22} & & S_{2n} \\ \vdots & & & \\ 0 & 0 & \cdots & S_{nn} \end{pmatrix}. \tag{5.21}$$

By the laws of matrix multiplication we obtain, from Eq. (5.21),

$$\begin{aligned} B_{ii} &= S_{1i}^2 + S_{2i}^2 + \cdots + S_{ii}^2, & i = j, \\ B_{ij} &= S_{1i}S_{1j} + S_{2i}S_{2j} + \cdots + S_{ii}S_{ij}. & i < j. \end{aligned} \tag{5.22}$$

From this pair of equations we may determine all the elements of $\mathbf{S}$. For the first row we have

$$S_{11} = \sqrt{B_{11}}, \qquad S_{1j} = \frac{B_{1j}}{S_{11}}. \tag{5.23a}$$

For succeeding rows, if $i > j$, $S_{ij} = 0$. Otherwise, if $j \geq i$,

$$\begin{aligned} S_{ii} &= \left(B_{ii} - \sum_{k=1}^{i-1} S_{ki}^2 \right)^{1/2}, & j = i, \\ S_{ij} &= \frac{B_{ij} - \sum_{k=1}^{i-1} S_{ki}S_{kj}}{S_{ii}}, & j > 1. \end{aligned} \tag{5.23b}$$

With Cholesky's method it is unnecessary to search for pivots. If we recall the discussion of Gaussian elimination in Chapter 4 we remember that a search for pivots was essential, because a zero could develop on the diagonal of the matrix $\mathbf{U}$ of the LU decomposition. But because $\mathbf{S}^T = \mathbf{L} \cdot \mathbf{D}^{1/2}$ and the d_{ii} are positive, so are the S_{ii}. Pivoting was also essential to prevent a division by a small element, with attendant accuracy loss. But the first part of Eq. (5.22) shows that

$$|S_{ij}| \leq \sqrt{B_{ii}}.$$

The elements of $\mathbf{S}$ cannot increase without limit even if pivoting is unused. That we can obviate a search for pivots was, if the reader recalls, of crucial importance for certain sparse matrix technqiues mentioned in Chapter 3.

As an example of a Cholesky decomposition let us calculate $\mathbf{S}$ for the positive definite 3×3 matrix

$$\mathbf{B} = \begin{pmatrix} 5 & 15 & 55 \\ 15 & 55 & 225 \\ 55 & 225 & 979 \end{pmatrix}. \tag{5.24}$$

We will work to five decimals, performing the calculations on a Sharp EL-5500 II pocket computer/calculator.

For the first row of $\mathbf{S}$

$$S_{11} = 2.23607, \qquad S_{12} = \frac{15}{2.23607} = 6.70820,$$

$$S_{13} = \frac{55}{2.23607} = 24.59673.$$

For the second row, $i = 2$, we find

$$S_{22} = (55 - 6.70820^2)^{1/2} = 3.16229$$

and

$$S_{23} = \frac{225 - 6.70820 \cdot 24.59673}{3.16229} = 18.97366.$$

Finally, for the last row

$$S_{33} = (979 - 24.59673^2 - 18.97366^2)^{1/2} = 3.74180.$$

Therefore,

$$\mathbf{S} = \begin{pmatrix} 2.23607 & 6.70820 & 24.59673 \\ 0 & 3.16229 & 18.97366 \\ 0 & 0 & 3.74180 \end{pmatrix}.$$

The elements of $\mathbf{S}$ are calculated with no difficulty despite $\mathbf{B}$'s having a relatively high condition number, 10^4 in the L_1 norm.

To illustrate the importance of positive definiteness, consider trying to decompose a symmetric but not positive definite matrix, namely

$$\mathbf{B} = \begin{pmatrix} 1 & 2 & 3 \\ 2 & 4 & 5 \\ 3 & 5 & 6 \end{pmatrix}. \tag{5.25}$$

This matrix has a lower condition number, ninety-eight in the L_1 norm, than the matrix (5.24). The calculation of the first row goes easily enough, giving $S_{11} = 1$, $S_{12} = 2$, $S_{13} = 3$. But for the second row we have

$$S_{22} = (4 - 2^2)^{1/2} = 0$$

and

$$S_{23} = \frac{5 - 2 \cdot 3}{0}.$$

S_{23} blows up because of the same difficulty that occurs with Gaussian elimination, developing a zero, S_{22} in this case, on the main diagonal. A permutation of the rows may cure such a problem, but we lose the main advantage of Cholesky decomposition, not having to search for pivots and interchange rows or columns. To continue this example, if we interchange the second and third rows we avoid a division by zero. The first row of S remains the same, and the second row becomes

$$S_{22} = (5 - 2^2)^{1/2} = 1$$

and

$$S_{23} = \frac{6 - 2 \cdot 3}{1} = 0.$$

Finally,

$$S_{33} = (5 - 3^2 - 0^2)^{1/2} = \sqrt{-1},$$

a complex number. Not only do we have to interchange rows, we also must use complex arithmatic. In summary, unless the matrix is positive definite, a Cholesky decomposition offers no advantage.

Having the Cholesky decomposition we still must solve the linear system

$$\mathbf{S}^T \cdot \mathbf{S} \cdot \mathbf{X} = \mathbf{A}^T \cdot \mathbf{d} = \mathbf{b}. \tag{5.26}$$

The easiest way to do this is to solve two triangular systems:

$$\begin{aligned} \mathbf{S} \cdot \mathbf{X} &= \mathbf{y}; \\ \mathbf{S}^T \cdot \mathbf{y} &= \mathbf{b}. \end{aligned} \tag{5.27}$$

Writing out in full the second part of Eq. (5.27)

$$\begin{pmatrix} S_{11} & 0 & \cdots & 0 \\ S_{12} & S_{22} & & 0 \\ \vdots & \vdots & & \vdots \\ S_{1n} & S_{2n} & & S_{nn} \end{pmatrix} \begin{pmatrix} y_1 \\ y_2 \\ \vdots \\ y_n \end{pmatrix} = \begin{pmatrix} b_1 \\ b_2 \\ \vdots \\ b_n \end{pmatrix}$$

allows us to derive

$$y_1 = \frac{b_1}{S_{11}}, \qquad i = 1,$$

$$y_i = \frac{b_i - \sum_{k=1}^{i-1} S_{ki} y_i}{S_{ii}}, \qquad i > 1. \tag{5.28}$$

Likewise, from the first part of Eq. (5.26), we have

$$X_n = \frac{y_n}{S_{nn}}, \qquad i = n,$$

$$X_i = \frac{y_i - \sum_{k=i+1}^{n} S_{ik} X_i}{S_{ii}}, \qquad i < n. \tag{5.29}$$

Going back to the matrix (5.24) suppose that the right-hand side is $\mathbf{b}^{\mathsf{T}} = (1 \quad 2 \quad 3)$. For y_1 we calculate $1/2.23607 = 0.44721$. Then

$$y_2 = \frac{2 - 6.70820 \cdot 0.44721}{3.16229} = -0.31622$$

and

$$y_3 = \frac{3 - 24.59673 \cdot 0.44721 + 18.97366 \cdot 0.31622}{3.74180}$$

$$= -0.53452.$$

Our system becomes

$$\begin{pmatrix} 2.23607 & 6.70820 & 24.59673 \\ 0 & 3.16229 & 18.97366 \\ 0 & 0 & 3.74180 \end{pmatrix} \begin{pmatrix} x_1 \\ x_2 \\ x_3 \end{pmatrix} = \begin{pmatrix} 0.44721 \\ -0.31622 \\ -0.53452 \end{pmatrix}$$

with the solution, calculated by Eq. (5.29), $\mathbf{X}^{\mathsf{T}} = (-0.49995 \quad 0.75710 \quad -0.14285)$. Had we used higher-precision arithmetic the correct solution to five decimals would be $\mathbf{X}^{\mathsf{T}} = (-0.50000 \quad 0.75714 \quad -0.14286)$. We have lost one digit in the process of solving the system, reasonable considering the number of arithmetic operations.

How many operations? Looking at Eq. (5.23b) we appreciate that the most frequent operation is the multiplication of S_{ki} by S_{kj}. This is buried in a loop running from 1 to $i-1$ contained in a loop running from $i+1$ to n (the loop for j) further contained in a loop running from 1 to n. The multplication will be executed

$$\sum_{i=1}^{n} \sum_{j=i+1}^{n} \sum_{k=1}^{i-1} 1$$

times. Expanding this expression, keeping only cubic terms as these control the operation count for $n \gg 1$, we find

$$\sum_{i=1}^{n} \sum_{j=i+1}^{n} (i-1) \cong \sum_{i=1}^{n} \left(\sum_{j=1}^{i} i - \sum_{j=1}^{i} i \right) = \sum_{i=1}^{n} (n_i - i^2)$$

$$= \frac{n^2(n+1)}{2} - \frac{n(n+1)(2n+1)}{6} \cong \frac{n^3}{2} - \frac{n^3}{3} = \frac{n^3}{6}.$$

The first part of Eq. (5.22) will be executed on the order of n^2 times, insignificant compared with the $n^3/6$ of the second part of Eq. (5.22), although the square root must be extracted n times. The solution of the two triangular systems also involves operations proportional to n^2. In general, the operation count is of the order of $n^3/6$.

In Section 4.4 it was mentioned that Gaussian elimination requires about $n^3/3$ operations. Where does Cholesky decomposition realize a savings of half the labor? By taking advantage of the symmetry of the matrix and cutting the number of operations in half. The dubious reader should study the coding for

Gaussian elimination, such as Chapter 17 of Forsythe and Moler's (1967) book. Such coding is not included here because Gaussian elimination is of only tangential interest to the main theme of this book. He will notice that the inner core of Gaussian elimination, like Cholesky decomposition, involves a multiplication buried within three nested DO loops:

$$\sum_{k=1}^{n-1} \sum_{i=k+1}^{n} \sum_{j=k+1}^{n} 1.$$

Upon expansion, and retaining only cubic terms, we arrive at $n^3/3$ for the operation count.

But Cholesky decomposition does necessitate extracting n square roots. It is possible, however, to perform the decomposition without having to calculate square roots. To see how, refer back to Eq. (5.19). To obtain the decomposition, the diagonal matrix $\mathbf{D}$ was absorbed into the triangular matrix $\mathbf{S}$, which no longer has unit elements on the main diagonal like $\mathbf{L}$. Rewrite Eq. (5.20) as

$$\mathbf{B} = \mathbf{G}^T \cdot \mathbf{D} \cdot \mathbf{G}, \tag{5.30}$$

where, once again, $\mathbf{G}$ is a triangular matrix with unit elements on the main diagonal.

Writing out Eq. (5.30) explicitly,

$$\begin{pmatrix} B_{11} & B_{12} & \cdots & B_{1n} \\ B_{21} & B_{22} & \cdots & B_{2n} \\ \vdots & \vdots & & \vdots \\ B_{n1} & B_{n2} & \cdots & B_{nn} \end{pmatrix} = \begin{pmatrix} 1 & 0 & \cdots & 0 \\ G_{12} & 1 & & 0 \\ \vdots & \vdots & & \vdots \\ G_{1n} & G_{2n} & \cdots & G_{nn} \end{pmatrix} \begin{pmatrix} d_1 & 0 & \cdots & 0 \\ 0 & d_2 & & 0 \\ \vdots & \vdots & & \vdots \\ 0 & 0 & \cdots & d_n \end{pmatrix}.$$

$$\begin{pmatrix} 1 & G_{12} & \cdots & G_{1n} \\ 0 & 1 & & G_{2n} \\ \vdots & \vdots & & \vdots \\ 0 & 0 & \cdots & G_{nn} \end{pmatrix}$$

$$= \begin{pmatrix} d_1 & 0 & \cdots & 0 \\ d_1 G_{12} & d_2 & & 0 \\ \vdots & \vdots & & \vdots \\ d_1 G_{1n} & d_2 G_{2n} & \cdots & d_n G_{nn} \end{pmatrix} \begin{pmatrix} 1 & G_{12} & \cdots & G_{1n} \\ 0 & 1 & & G_{2n} \\ \vdots & \vdots & & \vdots \\ 0 & 0 & & G_{nn} \end{pmatrix},$$

we deduce, for the first row of $\mathbf{G}$ and $\mathbf{D}$,

$$d_1 = B_{11}, \qquad G_{ij} = \frac{B_{1j}}{d_1}, \tag{5.31a}$$

and for succeeding rows

$$d_i = B_{ii} - \sum_{k=1}^{i-1} d_k G_{ki}^2, \qquad i > 1,$$

$$G_{ij} = \frac{B_{ij} - \sum_{k=1}^{i=1} d_k G_{ki} G_{kj}}{d_i}, \qquad j > 1. \tag{5.31b}$$

No square roots are involved, but the factorization is somewhat more complicated.

As an example take once more the matrix (5.24). From Eq. (5.31a) we calculate

$$d_1 = 5, \qquad G_{12} = \frac{15}{5} = 3, \qquad G_{13} = \frac{55}{5} = 11.$$

For the second and third rows from Eq. (5.31b)

$$d_2 = 55 - 5 \cdot 3^2 = 10, \qquad G_{23} = \frac{225 - 5 \cdot 3 \cdot 11}{10} = 6$$

and

$$d = (979 - 5 \cdot 11^2 - 10 \cdot 6^2) = 14.$$

Therefore,

$$\mathbf{B} = \begin{pmatrix} 1 & 0 & 0 \\ 3 & 1 & 0 \\ 11 & 6 & 1 \end{pmatrix} \begin{pmatrix} 5 & 0 & 0 \\ 0 & 10 & 0 \\ 0 & 0 & 14 \end{pmatrix} \begin{pmatrix} 1 & 3 & 11 \\ 0 & 1 & 6 \\ 0 & 0 & 1 \end{pmatrix}.$$

Which of the two Cholesky decompositions, what might be called the classical version or the square root-free version, is preferable? The number of arithmetic operations is about the same for both. The calculation of the square roots in the classical version may appear time-consuming, but the author performed timings of both versions (on a VAX-11/780 with 3 MB of physical memory and coding in FORTRAN-77) using a 100 × 100 matrix of double-precision numbers and found no significant difference between the two. Sometimes the square root-free version was marginally faster and sometimes the classical version was faster. And both were three times faster than Gaussian elimination. The factor of three, rather than two, as we might expect from a comparison of the operation counts, was undoubtedly due to the Gaussian elimination using subroutine calls. Here we have, once again, an instance of following supposed good programming practice, modularizing a program rather than creating a monolithic program, and being repaid by inefficiency as a result. The classical Cholesky decomposition, therefore, pays no severe timing penalty compared with its square root-free version. Because the factorization is more convenient—we need not worry about special coding to handle the elements of D—the classical Cholesky decomposition would generally be preferred.

5.3. The Variance–Covariance and Correlation Matrices

Although Chapter 2 recommended against inverting a matrix if our sole interest is solving a linear system, sometimes the inverse is desired in its own right. This is true in least squares problems, where a scaled inverse of the normal equations, called the variance–covariance matrix—often just covariance matrix, for short—provides a means of estimating the errors of the solution vector X and the correlations among the unknowns.

To understand why, we start with some assumptions. We assume that the errors in X depend only on d (see Eq. (5.15)), and that the errors in d are independent and follow the normal frequency distribution, Eq. (4.9). Let ε_j be the error in X caused by the jth component of d. From elementary calculus

$$\varepsilon_j = \frac{\partial X}{\partial d_j} \Delta d_j. \tag{5.32}$$

The total error in X, under the assumption of a normal distribution, is

$$\varepsilon^2 = \varepsilon^2 I = \sum_{j=1}^{m} \varepsilon_j \cdot \varepsilon_j^T. \tag{5.33}$$

From Eqns. (5.32) and (5.33) we have

$$\varepsilon^2 = \sum_{j=1}^{m} \frac{\partial X}{\partial d_j} \Delta d_j \left(\frac{\partial X}{\partial d_j} \Delta d_j \right)^T$$

$$= \sum_{j=1}^{m} \Delta d_j^2 \frac{\partial X}{\partial d_j} \left(\frac{\partial X}{\partial d_j} \right)^T \tag{5.34}$$

because Δd_j is a scalar. For the error Δd_j let us take a constant quantity, the mean error of unit weight $\sigma(1)$. If there are m equations of condition and n unknowns $\sigma(1)$, there σ of Eq. (4.9), is defined as

$$\sigma^2(1) = r^T \cdot \frac{r}{m-n}. \tag{5.35}$$

The problem now is to determine the partial derivatives $\partial X / \partial d_j$. From Eq. (5.15) we find that

$$\frac{\partial X}{\partial d_j} = (A^T \cdot A)^{-1} \cdot A_j^T, \tag{5.36}$$

where A_j^T is the jth column of A^T. Combining Eqns. (5.34), (5.35), and (5.36) we obtain

$$\varepsilon^2 = \sigma^2(1) \sum_{j=1}^{m} (A^T \cdot A)^{-T} \cdot A_j^T \cdot A_j \cdot (A^T \cdot A)^{-1}. \tag{5.37}$$

$A^T \cdot A$ is a symmetric matrix and, therefore,

$$(A^T \cdot A)^{-T} = (A^T \cdot A)^{-1}.$$

By the associativity of matrix multiplication

$$\varepsilon^2 = \sigma^2(1) \sum_{j=1}^{m} (A^T \cdot A)^{-1} \cdot A_j^T \cdot A_j \cdot (A^T \cdot A)^{-1}$$

$$= \sigma^2(1)(A^T \cdot A)^{-1} \sum_{j=1}^{m} A_j^T \cdot A_j \cdot (A^T \cdot A)^{-1}, \tag{5.38}$$

because $(A^T \cdot A)^{-1}$ is independent of j.

Looking carefully at the sum we see that it is a sum of the backwards product of $\mathbf{A}_j$ with its transpose,

$$\sum_{j=1}^{m} \mathbf{A}_j^{\mathrm{T}} \cdot \mathbf{A}_j = \sum_{j=1}^{m} \begin{pmatrix} a_{j1} \\ a_{j2} \\ \vdots \\ a_{jn} \end{pmatrix} (a_{j1} \quad a_{j2} \quad \cdots \quad a_{jn}), \tag{5.39}$$

which is nothing more than $\mathbf{A}^{\mathrm{T}} \cdot \mathbf{A}$. We arrive at

$$\varepsilon^2 = \varepsilon^2 \mathbf{I} = \sigma^2(1)(\mathbf{A}^{\mathrm{T}} \cdot \mathbf{A})^{-1}. \tag{5.40}$$

We define the unscaled covariance matrix as

$$\mathbf{C} = (\mathbf{A}^{\mathrm{T}} \cdot \mathbf{A})^{-1} = \mathbf{B}^{-1}, \tag{5.41}$$

the inverse of the matrix of the normal equations. The diagonal elements of $\mathbf{C}$ are the variances of the components of $\mathbf{X}$, and the off-diagonal elements are the covariances. Equation (5.40) shows that the diagonal elements of $\mathbf{C}$ when scaled by $\sigma^2(1)$ give the variances of the components of $\mathbf{X}$.

The correlation matrix is obtained from $\mathbf{C}$ by pre- and postmultiplying it by a diagonal matrix $\mathbf{D}$,

$$\mathbf{C}' = \mathbf{D} \cdot \mathbf{C} \cdot \mathbf{D}, \tag{5.42}$$

where $\mathbf{C}'$ is the correlation matrix and the elements of $\mathbf{D}$ are

$$\mathbf{D} = \begin{pmatrix} 1/\sqrt{C_{11}} & 0 & 0 \\ 0 & 1/\sqrt{C_{22}} & 0 \\ \vdots & \vdots & \vdots \\ 0 & 0 & 1/\sqrt{C_{nn}} \end{pmatrix}. \tag{5.43}$$

The elements of $\mathbf{C}'$ vary from zero to one, or zero to minus one. Zero means no correlation between variable i and variable j. One is perfect correlation, and minus one is perfect negative correlation.

Because the covariance and correlation matrices are so useful, we usually pay the computational price of inverting the matrix $\mathbf{B}$ rather than just solving a linear system. But because $\mathbf{C}$ and $\mathbf{C}'$ are so useful, and invariably exhibited in a least squares solution, we may easily lose sight of the assumptions on which they are based. To recapitulate, we have assumed that the errors in $\mathbf{X}$ depend only on $\mathbf{d}$, that the errors in $\mathbf{d}$ are independent and normally distributed, and that the standard derivation of the error Δd_j is constant.

How realistic are these assumptions? That the error in $\mathbf{X}$ depends only on $\mathbf{d}$ is defensible, as outlined in Chapter 4. That the errors in $\mathbf{d}$ are independent seems reasonable, but in many instances may not be as close to reality as we would like. Consider the observational errors in positions of minor planets, something with which the author has practical experience. If more than one minor planet is observed on a single photographic plate, then the errors in the positions will not be completely independent but rather correlated. The correlation depends on how many reference stars are common with the minor planets. Similar arguments apply in many disciplines, leaving the assumption of independence often more of a desire than a realistic postulate.

The assumption that the errors are normally distributed is, as remarked in Chapter 4, sometimes sound and sometimes far from the mark. That the standard deviation of the error Δd_j is constant is probably the least defensible assumption of all. For an experiment made by the same instrument during a relatively brief period a constant $\sigma(1)$ is no doubt a sound assumption. But for many experiments, where a number of instruments are used over a long period of time, $\sigma(1)$ undoubtedly varies with time and depends also on the instrument used.

If $\sigma(1)$ is a function of time nothing prevents us from leaving Δd_j within the summation of Eq. (5.34). In such a case we arrive at, instead of Eq. (5.38),

$$\varepsilon_i^2 = (\mathbf{A}^T \cdot \mathbf{A})^{-1} \sum_{j=1}^{m} \Delta d_j^2 \, \mathbf{A}_j^T \cdot \mathbf{A}_j \cdot (\mathbf{A}^T \cdot \mathbf{A})^{-1}. \tag{5.44}$$

Each of the backwards products $\mathbf{A}_j^T \cdot \mathbf{A}_j$ is weighted by Δd_j^2. We may still arrive at error estimates for the components of $\mathbf{X}$ and correlations among the componnts, but the procedure is more complicated than merely inverting a matrix and associating its scaled diagonal elements with variances.

Unless the assumptions upon which Eq. (5.40) are realized in practice, the error estimates for $\mathbf{X}$ and the correlations from Eq. (5.42) are likely to be more formalities than genuine error estimates and correlations. But assuming that we really do want to use Eq. (5.40), the inverse $\mathbf{B}^{-1}$ is easily calculated from Eq. (3.5) of Chapter 3. To reiterate what was mentioned there these equations permit a matrix, in this case a Cholesky factor $\mathbf{S}$, to be inverted in-place in the space already allocated for $\mathbf{S}$.

Cholesky decomposition requires on the order of $n^3/6$ arithmetic operations. To find how many we need to invert $\mathbf{S}$, look at Eq. (3.5). We have three nested DO loops with a multiplication buried inside the innermost loop. We need.

$$\sum_{i=1}^{n-1} \sum_{j=i+1}^{n} \sum_{k=1}^{j-1} 1$$

operations. Keeping only terms of order three we find that this expands to $n^3/6$ operations. For the Cholesky decomposition of $\mathbf{S}$ and the calculation of $\mathbf{S}^{-1}$ we require altogether $n^3/3$ operations, still substantially less than the n^3 needed for the inverse of a general matrix.

As an example let us consider the inverse of the Cholesky factor of matrix (5.24). Because Eq. (3.5) is so simple we present a few lines of BASIC code to invert S;

```
100   FOR I = 1 TO N
110   FOR J = I TO N
120   IF (J < > I) GO TO 150
130   A(I, I) = 1.0/A(I, I)
140   GO TO 200
150   SUM = 0.0
160   FOR K = I TO (J − 1)
170   SUM = SUM + A(I, K) * A(K, J)
180   NEXT K
190   A(I, J) = −SUM/A(J, J)
200   NEXT J
210   NEXT I
```

The author used this coding with his Sharp EL-5500 II pocket calculator/ computer to calculate S^{-1}, given here to five rather than ten decimals, as

$$S^{-1} = \begin{pmatrix} 0.44721 & -0.94868 & 1.87073 \\ 0 & 0.31623 & -1.60350 \\ 0 & 0 & 0.26725 \end{pmatrix}.$$

B^{-1} is found from $S^{-1} \cdot S^{-T}$ and is

$$B^{-1} = \begin{pmatrix} 4.59962 & -3.29972 & 0.49995 \\ -3.29972 & 2.67121 & -0.42854 \\ 0.49995 & -0.42854 & 0.07142 \end{pmatrix}.$$

B^{-1} is the unscaled covariance matrix. The correlation matrix follows from Eq. (5.42) as

$$C' = \begin{pmatrix} 0.46627 & 0 & 0 \\ 0 & 0.61185 & 0 \\ 0 & 0 & 3.74188 \end{pmatrix}$$

$$\cdot B^{-1} \cdot \begin{pmatrix} 0.46627 & 0 & 0 \\ 0 & 0.61185 & 0 \\ 0 & 0 & 3.74188 \end{pmatrix}$$

$$= \begin{pmatrix} 1.00000 & -0.94137 & 0.87228 \\ -0.94137 & 1.00000 & -0.98113 \\ 0.87228 & -0.98113 & 1.00000 \end{pmatrix}.$$

All of the unknowns are highly correlated. In actual practice we would not pre- and postmultiply C by D, convenient notationally but inefficient computationally. Rather, each element C'_{ij} of C' is calculated by

$$C'_{ij} = \frac{C_{ij}}{\sqrt{C_{ii}}\sqrt{C_{jj}}}. \tag{5.45}$$

The diagonal elements are all unity because an unknown is perfectly correlated with itself and need not be calculated. Because of symmetry only the upper (or lower) off-diagonal elements must be calculated.

If we use the square root-free version of the Cholesky decomposition, Eqns. (3.5) are still valid, but it would be better to revise them slightly to take advantage of the unit entries on the diagonal of **G**. These unit entries need not be stored at all and may be replaced by the elements of the diagonal matrix **D**. The following BASIC code uses an $n \times n$ array for both the elements of **G** and **D**.

```
100   FOR I = 1 TO N
110   G(I, I) = 1.0/G(I, I)
120   FOR J = (I + 1) TO N
130   SUM = 0.0
140   FOR K = I TO (J − 1)
150   IF (I = K) THEN GIK = 1.0
160   IF (I < > K) THEN GIK = G(I, K)
170   SUM = SUM + GIK * G(K, J)
180   NEXT K
190   G(I, J) = −SUM
200   NEXT J
210   NEXT I
```

To calculate $\sigma(1)$, the mean error of unit weight, we must, by Eq. (5.35), feed the solution **X** into the equations of condition and calculate **r**. It is, in general, a good idea to have the individual residuals because a large residual flags a discordant observation that must subsequently be eliminated; the individual residuals are also necessary for iteratively reweighted least squares, discussed in Section 5.5. But we may also find $\sigma(1)$ without explicitly calculating the residuals, which may occasionally be convenient, especially after discordant observations have been discarded. From Eq. (5.35)

$$\sigma^2(1) = (\mathbf{A} \cdot \mathbf{X} - \mathbf{d})^\mathrm{T} \cdot \frac{\mathbf{A} \cdot \mathbf{X} - \mathbf{d}}{m - n}. \tag{5.46}$$

Considering only the numerator of Eq. (5.46) we find

$$(\mathbf{A} \cdot \mathbf{X} - \mathbf{d})^\mathrm{T} \cdot (\mathbf{A} \cdot \mathbf{X} - \mathbf{d}) = (\mathbf{X}^\mathrm{T} \cdot \mathbf{A}^\mathrm{T} - \mathbf{d}^\mathrm{T}) \cdot (\mathbf{A} \cdot \mathbf{X} - \mathbf{d})$$

$$= \mathbf{X}^\mathrm{T} \cdot \mathbf{A}^\mathrm{T} \cdot \mathbf{A} \cdot \mathbf{X} - \mathbf{d}^\mathrm{T} \cdot \mathbf{A} \cdot \mathbf{X} - \mathbf{X}^\mathrm{T} \cdot \mathbf{A}^\mathrm{T} \cdot \mathbf{d} + \mathbf{d}^\mathrm{T} \cdot \mathbf{d}$$

$$= \mathbf{X}^\mathrm{T} \cdot \mathbf{B} \cdot \mathbf{X} - (\mathbf{A} \cdot \mathbf{X})^\mathrm{T} \cdot \mathbf{d}^\mathrm{TT} - \mathbf{X}^\mathrm{T} \cdot \mathbf{A}^\mathrm{T} \cdot \mathbf{d} + \mathbf{d}^\mathrm{T} \cdot \mathbf{d}$$

$$= \mathbf{X}^\mathrm{T} \cdot \mathbf{B} \cdot \mathbf{X} - \mathbf{X}^\mathrm{T} \cdot \mathbf{A}^\mathrm{T} \cdot \mathbf{d} - \mathbf{X}^\mathrm{T} \cdot \mathbf{A}^\mathrm{T} \cdot \mathbf{d} + \mathbf{d}^\mathrm{T} \cdot \mathbf{d}$$

$$= \mathbf{X}^\mathrm{T} \cdot (\mathbf{B} \cdot \mathbf{X} - \mathbf{A}^\mathrm{T} \cdot \mathbf{d}) - \mathbf{X}^\mathrm{T} \cdot \mathbf{A}^\mathrm{T} \cdot \mathbf{d} + \mathbf{d}^\mathrm{T} \cdot \mathbf{d}$$

$$= \mathbf{d}^\mathrm{T} \cdot \mathbf{d} - \mathbf{X}^\mathrm{T} \cdot \mathbf{A}^\mathrm{T} \cdot \mathbf{d}.$$

Therefore,

$$\sigma^2(1) = \frac{\mathbf{d}^\mathrm{T} \cdot \mathbf{d} - \mathbf{X}^\mathrm{T} \cdot \mathbf{A}^\mathrm{T} \cdot \mathbf{d}}{m - n}. \tag{5.47}$$

One of the pleasant consequences of a solution to a least squares problem, by the formation of normal equations and Cholesky decomposition, follows from the symmetry of the equations. B is symmetric, requiring $n(n + 1)/2$ rather than n^2 locations, the upper triangular matrix S may be developed in this space, and S^{-1} may also be computed in the same space. Figure 5.1 gives FORTRAN coding for a linear least squares analysis, involving the formation of the normal equations, their Cholesky decomposition, the calculation of the covariance matrix, along with the errors in the solution vector, and the calculation of the correlation matrix. The program uses an array of size $n(n + 1)/2$ for the symmetric B matrix, the Cholesky factor S, and its inverse S^{-1}. Two additional arrays of size n for each one are used for the vectors X and Y of Section 5.2, and an array of size $n + 1$ for each of the equations of condition. $\sigma(1)$ is found from the explicit calculation of the residuals, which involves rereading the equations of condition. In other words, we must input the equations of condition at least twice. If we wish to avoid this and to employ Eq. (5.47), additional space would have to be allocated to contain $A^T \cdot d$, which gets overwritten in solving for Y and afterwards X, and for the scalar product $d^T \cdot d$. Nevertheless, we still save almost half the space, for large n, of what would be required if we used a two-dimensional array for B.

5.4. Orthogonal Transformations

The solution of the overdetermined system Eq. (4.1) by the method of least squares is usually done by use of the normal equations. But there are other ways to obtain a solution, ways that do not suffer from the drawback mentioned in Section 5.1 that, aside from special circumstances, the condition number of the matrix of the normal equations is the square of that of the equations of condition. The alternatives to normal equations involve orthogonal transformations.

In Chapter 2 we stated that orthogonal matrices are intimately involved with the method of least squares. Orthogonal matrices conserve the Euclidean norm of a vector. Let Q be an orthogonal matrix and A an arbitrary $n \times n$ matrix, nonsingular of course. Then

$$\|Q \cdot A\|_2 = (Q \cdot A)^T \cdot (Q \cdot A) = A^T \cdot Q^T \cdot Q \cdot A$$

$$= A^T \cdot A = \|A\|_2. \tag{5.48}$$

Orthogonal matrices possess unit condition number in the L_2 norm. From Eq. (2.28)

$$\|Q\|_2 = \lambda_{max}^{1/2},$$

where λ_{max} is the largest eigenvalue of $Q^T \cdot Q$. But this product is just I and, therefore, $\lambda_{max} = 1$. Likewise, as discussed in Chapter 2,

$$\|Q^{-1}\|_2 = \|Q^T\|_2 = \lambda_{min}^{1/2}.$$

But, once again, $\mathbf{Q} \cdot \mathbf{Q}^T = \mathbf{I}$ and $\lambda_{min} = 1$. The ratio $\lambda_{max}/\lambda_{min}$ is unity. If, therefore, a matrix has a given condition number, multiplying it by an orthogonal matrix does not increase its condition number in the L_2 norm.

For the least squares problem we have, from Eq. (5.48),

$$\|\mathbf{A} \cdot \mathbf{X} - \mathbf{d}\|_2 = \|\mathbf{Q} \cdot \mathbf{A} \cdot \mathbf{X} - \mathbf{Q} \cdot \mathbf{d}\|_2 = \min. \qquad (5.49)$$

If we find some matrix $\mathbf{Q}$ that satisfies Eq. (5.49) we will obtain a least squares solution without having formed normal equations with their higher condition number.

Suppose that we find a $\mathbf{Q}$ such that

$$\mathbf{Q} \cdot \mathbf{A} = \mathbf{R}, \qquad (5.50)$$

where $\mathbf{R}$ is $m \times n$ with the lower $(m - n) \times n$ elements null and the upper $n \times n$ elements forming an upper triangular matrix. $\mathbf{Q}$ must, evidently, be of size $m \times m$. $\mathbf{R}$ is upper triangular and so is $\mathbf{S}$ of the Cholesky decomposition of $\mathbf{B}$. Could there be a relation between the two?

Going back to Eq. (5.1), multiplying both sides by $\mathbf{Q}$, and following closely the development of Eqns. (5.2)–(5.6) we have

$$\mathbf{Q} \cdot \mathbf{A} \cdot \mathbf{X} = \mathbf{R} \cdot \mathbf{X} = \mathbf{Q} \cdot \mathbf{d} + \mathbf{Q} \cdot \mathbf{r}. \qquad (5.51)$$

Taking differentials of both sides we obtain

$$\mathbf{R} \cdot d\mathbf{X} = \mathbf{Q} \cdot d\mathbf{r}, \qquad (5.52)$$

and from $d\mathbf{r}^T \cdot \mathbf{r} = 0$ we get

$$(\mathbf{Q}^T \cdot \mathbf{R} \cdot d\mathbf{X})^T \cdot \mathbf{r} = 0, \qquad (5.53)$$

or

$$d\mathbf{X}^T \cdot \mathbf{R}^T \cdot \mathbf{Q} \cdot (\mathbf{Q}^T \cdot \mathbf{R} \cdot \mathbf{X} - \mathbf{d}) = 0. \qquad (5.54)$$

By the associativity of matrix multiplication

$$d\mathbf{X}^T \cdot \mathbf{R}^T \cdot (\mathbf{R} \cdot \mathbf{X} - \mathbf{Q} \cdot \mathbf{d}) = 0. \qquad (5.55)$$

Because $d\mathbf{X}^T$ is an arbitrary vector we infer that, in general,

$$\mathbf{R} \cdot \mathbf{X} = \mathbf{Q} \cdot \mathbf{d}. \qquad (5.56)$$

By comparing Eq. (5.56) with Eq. (5.6) and using Eq. (5.20) we see that

$$\mathbf{R} = \mathbf{S} \qquad (5.57)$$

and

$$\mathbf{Q} \cdot \mathbf{d} = \mathbf{S}^{-T} \cdot \mathbf{A}^T \cdot \mathbf{d}. \qquad (5.58)$$

We can, with an orthogonal transformation, arrive at the same set of equations as those given by a Cholesky decomposition of the normal equations, but without working with a matrix of higher condition number than $\mathbf{A}$. This fact causes some authors, such as Lawson and Hanson (1974, p. 122), to state flatly that orthogonal transformations handle a least squares problem in single-precision, whereas normal equations, because of their higher condition number, require double-precision. But this assertion, as we shall see shortly, is too simplistic.

The matrix $\mathbf{Q}$ is built up by multiplication of a series of elementary orthogonal matrices. Two matrices are commonly used: the Givens rotation matrix (Givens, 1954), and the Householder reflection matrix (Householder, 1958).

The Givens matrix is nothing more than a rotation matrix,

$$\mathbf{Q} = \mathbf{G} = \begin{pmatrix} \cos\theta & \sin\theta \\ -\sin\theta & \cos\theta \end{pmatrix}, \tag{5.59}$$

with θ defined so that when $\mathbf{G}$ is multiplied into a two-vector an element of the vector is zeroed,

$$\mathbf{G} \cdot \begin{pmatrix} X_1 \\ X_2 \end{pmatrix} = \begin{pmatrix} r \\ 0 \end{pmatrix}. \tag{5.60}$$

From Eqns. (5.59) and (5.60) we deduce

$$\begin{aligned} \cos\theta X_1 + \sin\theta X_2 &= r, \\ -\sin\theta X_1 + \cos\theta X_2 &= 0, \end{aligned} \tag{5.61}$$

from which, from the second equation of this pair,

$$\theta = \tan^{-1}\left(\frac{X_1}{X_2}\right). \tag{5.62}$$

Rather than use the arctangent it is more efficient computationally to express the sine and cosine in Eq. (5.61) in terms of square roots. Solving Eq. (5.61) for $\cos\theta$ and $\sin\theta$ in terms of X_1 and X_2, we find

$$r = (X_1^2 + X_2^2)^{1/2},$$

$$\cos\theta = \frac{X_1}{r}, \qquad \sin\theta = \frac{X_2}{r}. \tag{5.63}$$

To apply the Givens rotations to an $m \times n$ matrix $\mathbf{A}$ we modify an $m \times m$ unit matrix by placing the sine and cosine of the rotation in a position to zero one element of $\mathbf{A}$. Consider the linear system

$$\begin{pmatrix} 1 & 1 \\ 1 & 2 \\ 1 & 3 \\ 1 & 4 \\ 1 & 5 \end{pmatrix} \begin{pmatrix} X_1 \\ X_2 \end{pmatrix} = \begin{pmatrix} 1 \\ 1 \\ 2 \\ 3 \\ 2 \end{pmatrix}. \tag{5.64}$$

To zero the element A_{51} we have, from Eq. (5.63), $r = (1^2 + 1^2)^{1/2}$, $\cos\theta = 1/\sqrt{2}$, $\sin\theta = 1/\sqrt{2}$. The Givens matrix will zero element A_{51} and alter element A_{41} and must, therefore, be a 5×5 unit matrix with elements G_{44},

G_{45}, G_{54}, G_{55} replaced by the 2×2 submatrix Eq. (5.59), or

$$
\mathbf{G} = \begin{pmatrix}
1 & 0 & 0 & 0 & 0 \\
0 & 1 & 0 & 0 & 0 \\
0 & 0 & 1 & 0 & 0 \\
0 & 0 & 0 & 1/\sqrt{2} & 1/\sqrt{2} \\
0 & 0 & 0 & -1/\sqrt{2} & 1/\sqrt{2}
\end{pmatrix}. \tag{5.65}
$$

After multiplication by $\mathbf{G}$, Eq. (5.64) becomes

$$
\begin{pmatrix}
1 & 1 \\
1 & 2 \\
1 & 3 \\
\sqrt{2} & 6.36396 \\
0 & 0.70711
\end{pmatrix}
\begin{pmatrix} X_1 \\ X_2 \end{pmatrix}
=
\begin{pmatrix}
1 \\
1 \\
2 \\
3.53553 \\
-0.70711
\end{pmatrix}.
$$

We would have to apply a total of seven Givens rotations to reduce Eq. (5.64) to upper triangular form. For an $m \times n$ matrix we need, in general, $mn - n(n+1)/2$ transformations to reduce an $m \times n$ matrix to upper triangular form.

Although it is notationally elegant to express $\mathbf{Q}$ as a series of multiplied elementary orthogonal transformations,

$$
\mathbf{Q} = \mathbf{G}_k \cdot \mathbf{G}_{k-1} \cdot \ldots \cdot \mathbf{G}_1,
$$

computationally such a procedure would be ridiculously inefficient. Each Givens matrix, although of size $m \times m$, alters only two elements in each column of $\mathbf{A}$ and in $\mathbf{d}$. Figure 5.2 exhibits the FORTAN code for the efficient implementation of the Givens rotations. The code implements the Givens rotations in such a way that no explicit matrix $\mathbf{G}$ is needed. To protect against possible overflow or underflow the program uses the trick

$$
\begin{aligned}
a_1 &= \max(|X_1|, |X_2|), \\
a_2 &= \min(|X_1|, |X_2|), \\
r &= a_1\sqrt{1 + (a_2/a_1)^2}.
\end{aligned} \tag{5.66}
$$

Although we reduce an $m \times n$ matrix to upper triangular form by orthogonal transformations and avoid the higher condition number of the normal equations, a careful study of Figure 5.2 should cause us to stop and reflect. To form normal equations takes, because the matrix $\mathbf{B}$ is symmetric, $mn^2/2$ operations. The Cholesky decomposition of the normal equations requires an additional $n^3/6$ operations. The operation count for the solution is, for reasonable sized m and n, negligible compared with the $mn^2/2 + n^3/6$ operations. Figure 5.2 shows a loop running from 1 to n, within that one running from $j + 1$ to m, and within that one running from j to n. The innermost DO loop involves four multiplications. Altogether, we need

$$\sum_{j=1}^{n} \sum_{i=j+1}^{m} \sum_{k=j}^{n} 4$$

operations. Expanding these sums and, as usual, keeping only terms proportional to cubic power we find an operation count of $2mn^2 - 2n^3/3$. This is, for $m \gg n$, about four times more work compared with the normal equation approach. Recalling the discussion of the statistical accumulation of error in Chapter 1, we realize that there is bound to be more error accumulation with the Givens transformations. Furthermore, we also extract $mn - n(n + 1)/2$ square roots compared with n for the Cholesky decomposition of the normal equations. These are disturbing thoughts. Perhaps in an excess of zeal to avoid higher condition numbers, we overlook a greater accumulation of error, rounding or truncation depending on the computer. We shall return to this consideration in the next section.

Just as there is a square root-free version of the Cholesky decomposition, so there is a square root-free version of the Givens orthogonal matrix, discovered by Gentleman and discussed in Lawson and Hanson (1974, pp. 60–62). The innermost loop of the square root-free Givens rotations involves only two, rather than four, multiplications, for a seeming substantial savings of time. But the Gentleman algorithm also requires an auxiliary array of size m to contain diagonal elements, and the coding is considerably more tricky than that of the standard Givens method. Hanson (1973) found that on a Univac 1108 the Gentleman version of the Givens method was only 25% faster than the standard version, rather than 50% faster as we might intuitively expect on the basis of two rather than four multiplications. In any event, we still do more work than with the formation and solution of normal equations.

Another frequently used orthogonal transformation, Householder's, also involves less work than the Givens transformation, about half as much. In his original paper Householder (1958) commented that the number of square roots required in the Givens triangularization of a matrix led him to seek a different transformation, one that turns out to be, if more efficient, also more complicated than a simple plane rotation.

Householder sought a symmetric, orthogonal matrix that would zero all of the elements of a vector except the first:

$$\mathbf{H} \cdot \mathbf{X} = k\mathbf{e}_1; \tag{5.67}$$

$\mathbf{e}_1^{\mathrm{T}} = (1 \ 0 \ 0 \ \ldots)$, and k is the Euclidean norm of $\mathbf{X}$. A matrix satisfying Eq. (5.67) is

$$\mathbf{H} = \mathbf{I} - 2\mathbf{w} \cdot \mathbf{w}^{\mathrm{T}} \qquad (\mathbf{w}^{\mathrm{T}} \cdot \mathbf{w} = 1). \tag{5.68}$$

$\mathbf{w}$ is a unitary vector to be determined. By definition its scalar product is unity, and $\mathbf{w} \cdot \mathbf{w}^{\mathrm{T}}$ is another example of the backwards product.

By definition H is symmetric and orthogonal. From Eq. (5.68)

$$\begin{aligned} H \cdot H &= (I - 2w \cdot w^T) \cdot (I - 2w \cdot w^T) \\ &= I - 4w \cdot w^T + 4w \cdot w^T \cdot w \cdot w^T \\ &= I = H^{-1} \cdot H. \end{aligned} \tag{5.69}$$

The Householder matrix is, therefore, its own inverse, a far from common matrix.

If X is an arbitrary m vector multiplying it by H zeroes all of the elements except the first. This is equivalent to applying $(m - 1)$ plane notations with one difference. The eigenvalues of H are real, whereas those of G are, in general, complex. The eigenvalue–eigenvector decomposition of H is

$$H = V \cdot \lambda \cdot V^T, \tag{5.70}$$

from which

$$\begin{aligned} I = H^T \cdot H &= V \cdot \lambda^T \cdot V^T \cdot V \cdot \lambda \cdot V^T \\ &= V \cdot \lambda^2 \cdot V^T \quad \Rightarrow \quad V^T \cdot I \cdot V = I = \lambda^2. \end{aligned} \tag{5.71}$$

The eigenvalues are real, but not all positive. In fact,

$$H \cdot w = w - 2w \cdot w^T \cdot w = -w, \tag{5.72}$$

or $\lambda = -1$.

All of this is a long-winded way of saying that H is a reflection matrix, rather than a rotation matrix like G; H reverses the orientation of the configuration. The Givens matrix may also be expressed as a reflection, rather than rotation, matrix. In symmetric form

$$G = \begin{pmatrix} \cos \theta & \sin \theta \\ \sin \theta & -\cos \theta \end{pmatrix}. \tag{5.73}$$

For $\theta = 0°$, the Givens rotation matrix reduces to a unit matrix and leaves unchanged a unit vector $X^T = (1/\sqrt{2} \ \ 1/\sqrt{2})$: the Givens reflection matrix for the same angle reduces to

$$\begin{pmatrix} 1 & 0 \\ 0 & -1 \end{pmatrix}$$

and reflects the unit vector about the X_1-axis. A 2×2 Householder matrix is nothing more than a Givens reflection matrix, as we shall see.

To determine the vector w in Eq. (5.68) we shall follow closely the reasoning in Householder's original article (1958). Combining Eqns. (5.67) and (5.68) we have

$$(I - 2w \cdot w^T) \cdot X = ke_1 \quad \Rightarrow \quad X - 2w(w^T \cdot X) = ke_1. \tag{5.74}$$

Denote $w^T \cdot X$, a scalar, by μ. Then

$$2w\mu = X - ke_1,$$

and upon multiplying both sides by their transpose we get

$$(2\mathbf{w}\mu)^T \cdot (2\mathbf{w}\mu) = (\mathbf{X} - k\mathbf{e}_1)^T \cdot (\mathbf{X} - k\mathbf{e}_1)$$

or

$$4\mathbf{w}^T \cdot \mathbf{w}\mu^2 = \mathbf{X}^T \cdot \mathbf{X} - 2k\mathbf{X}^T \cdot \mathbf{e}_1 + k^2.$$

But $\mathbf{w}^T \cdot \mathbf{w} = 1$ and, from Eq. (5.67),

$$(\mathbf{X}^T \cdot \mathbf{H}^T) \cdot (\mathbf{H} \cdot \mathbf{X}) = \mathbf{X}^T \cdot \mathbf{X} = k^2.$$

Therefore,

$$2\mu^2 = k^2 \mp kX_1 \qquad (5.75)$$

and

$$2\mathbf{w}\mu = \mathbf{X} \mp k\mathbf{e}_1. \qquad (5.76)$$

Equations (5.75) and (5.76) allow us to determine μ and w. The $\mp$ sign in Eq. (5.75) comes from $k^2 = \mathbf{X}^T \cdot \mathbf{X} = \mathbf{X}^2$ and is chosen to make $X_1 \mp k$ in the first component of Eq. (5.76) large in absolute value, which enhances numerical stability.

To illustrate the calculation of a Householder transformation we shall zero the elements except the first of the first column of Eq. (5.64). The vector $\mathbf{X}^T = (1\ \ 1\ \ 1\ \ 1\ \ 1)$ and $k^2 = \mathbf{X}^T \cdot \mathbf{X} = 5$. Thus,

$$k = \pm\sqrt{5}$$

and from Eq. (5.75), $2\mu^2 = 5 + \sqrt{5}$. Taking the plus, rather than the minus, sign makes $2\mu^2$ larger, 7.23607 rather than 2.76393. Then $\mu = 1.90211$ and, from Eq. (5.76),

$$\mathbf{w} = \frac{1}{3.80422} \begin{pmatrix} 1 + \sqrt{5} \\ 1 \\ 1 \\ 1 \\ 1 \end{pmatrix} = \begin{pmatrix} 0.85065 \\ 0.26287 \\ 0.26287 \\ 0.26287 \\ 0.26287 \end{pmatrix}.$$

$$\mathbf{w} \cdot \mathbf{w}^T = \begin{pmatrix} 0.72361 & 0.22361 & 0.22361 & 0.22361 & 0.22361 \\ 0.22361 & 0.06910 & 0.06910 & 0.06910 & 0.06910 \\ 0.22361 & 0.06910 & 0.06910 & 0.06910 & 0.06910 \\ 0.22361 & 0.06910 & 0.06910 & 0.06910 & 0.06910 \\ 0.22361 & 0.06910 & 0.06910 & 0.06910 & 0.06910 \end{pmatrix}.$$

Finally, from Eq. (5.68) we compute

$$\mathbf{H} = \begin{pmatrix} -0.44722 & -0.44722 & -0.44722 & -0.44722 & -0.44722 \\ -0.44722 & 0.86180 & -0.13820 & -0.13820 & -0.13820 \\ -0.44722 & -0.13820 & 0.86180 & -0.13820 & -0.13820 \\ -0.44722 & -0.13820 & -0.13820 & 0.86180 & -0.13820 \\ -0.44722 & -0.13820 & -0.13820 & -0.13820 & 0.86180 \end{pmatrix}.$$

Applying this matrix to $\mathbf{X}$ we find

$$\begin{pmatrix} -2.23610 \\ -0.00002 \\ -0.00002 \\ -0.00002 \\ -0.00002 \end{pmatrix}.$$

To the accuracy of the calculation the lower four elements of the five-vector are indeed zeroed. The reader can see immediately that the calculated $\mathbf{H}$ is symmetric. He should also verify some of the other relations proved here, for example, that $\mathbf{H} \cdot \mathbf{w} = -\mathbf{w}$ and $\mathbf{H}^{-1} = \mathbf{H}$. Seeing the numbers fall into place often has more impact than a dry mathematical proof.

As with the Givens transformation it is grossly inefficient to explicitly calculate the matrix $\mathbf{H}$. For an $m \times n$ least squares problem, $\mathbf{H}$ is of size $m \times m$, which can be huge for large m. Instead of compressing a matrix of size $m \times n$ to one of size $n \times n$, as with normal equations, we seem to be expanding it to one of size $m \times m$. And, unlike the situation with the Givens transformations, the Householder matrix is not sparse. Not sparse, but because $\mathbf{I}$ is known and $\mathbf{w} \cdot \mathbf{w}^T$ is of rank one, many of the m^2 elements are redundant. The entire transformation is defined by the m vector $\mathbf{X}$ and the scalar μ. In practice, we invariably apply $\mathbf{H}$ in factored form,

$$\mathbf{H} \cdot \mathbf{X} = \mathbf{X} - 2\mathbf{w} \cdot (\mathbf{w}^T \cdot \mathbf{X}) = \mathbf{X} - 2\mathbf{w}\mu. \tag{5.77}$$

Matrix notation, although indispensible for derivations and for exhibiting the structure of mathematical operations, is often computationally inefficient.

Given an $m \times n$ overdetermined system and a right-hand side, we must apply $n + 1$ Householder transformations to reduce the system to upper triangular form. Figure 5.3 shows FORTRAN code for the triangularization of the system. At no point do we work with an $m \times m$ matrix. The analysis of the complexity of the algorithm is somewhat more complicated than with the Givens transformation. For comparability with the earlier analyses of the Givens and normal equations complexity, we leave out the reduction of the right-hand side and consider only the trangularization of the $m \times n$ matrix.

Studying the DO loops and considering only multiplications and divisions, we see that

$$\sum_{j=1}^{n} \sum_{i=j}^{m} 1 + \sum_{j=1}^{n} \sum_{k=j}^{n} \sum_{i=j}^{m} 2$$

operations are needed. The double sum is the operation count for constructing the transformations. Expanding the sums and, as usual, keeping only cubic terms, we find that about $mn^2 - n^3/3$ operations are required. For large m this is twice fewer operations than with the Givens transformations but still twice as many as required by normal equations.

Where does the Householder transformation realize its efficiency compared with the Givens transformation? For a two-vector the two are equivalent. This may be shown easily enough from the relations already developed. From

$$\mathbf{H} \cdot \begin{pmatrix} X_1 \\ X_2 \end{pmatrix} = \begin{pmatrix} r \\ 0 \end{pmatrix} \tag{5.78}$$

we find $k^2 = X_1^2 + X_2^2 = r^2$, $2\mu^2 = r^2 + X_1 r$ (taking the plus sign for convenience), and

$$2\mathbf{w} \cdot \mathbf{w}^T = \begin{pmatrix} X_1/r - 1 & X_2/r \\ X_2/r & X_2^2/r(X_1 + r) \end{pmatrix}. \tag{5.79}$$

Upon subtracting Eq. (5.79) from a unit matrix and clearing fractions, we obtain

$$\mathbf{H} = \begin{pmatrix} -X_1/r & -X_2/r \\ -X_2/r & X_1/r \end{pmatrix}, \tag{5.80}$$

a Givens reflection matrix multiplied by -1.

For a two-vector, therefore, the efficiency of the two transformations is the same. But as m grows with respect to two, the Householder transformation does more of its work outside of the three-deep nested DO loop. The construction of the Householder transformations involves a two-deep nested DO loop, as does the construction of a Givens transformation, but the $\mathbf{H}$ matrix, being more complicated than the $\mathbf{G}$ matrix, requires only two multiplications within the three-deep nested DO loop compared with four for the latter. But both the Householder and the Givens transformations do more work than the formation of normal equations and their Cholesky decomposition, because the formation of the normal equations involves constant $mn^2/2$ operations. The arithmetic operations for the Cholesky decomposition involve a matrix of size $n \times n$, or $n(n + 1)/2$ if we take advantage of the symmetry. With orthogonal transformations, on the other hand, the arithmetic operations involve a matrix of size $m \times n$. The net effect is to multiply the constant $mn^2/2$ by a factor, four for the Givens transformations and two for Householder. We save somewhat on the operation count because we both reduce an $m \times n$ matrix and simultaneously triangularize it; this subtracts a term proportional to n^3 from the operation count. Once the normal equations have been formed they must be subsequently triangularized, which adds a term proportional to n^3 to the operation count.

The matrix $\mathbf{Q}$ that reduces the $m \times n$ matrix $\mathbf{A}$ to upper triangular form, whether it is built up by the Givens or Householder transformations, is unique. The reader who is not convinced of this by the somewhat heuristic discussion at the beginning of this section can find a more rigorous proof in Section 4.53 of Wilkinson's treatise (1965). It does not follow, however, that the orthogonal matrix, built up from the Givens transformations to zero the elements of a given column, will be the same as the Householder matrix to zero the same elements. The latter is both orthogonal and symmetric, but the product of the Givens rotations, while orthogonal, will not in general be symmetric.

To make this more concrete, consider the triangularization of

$$\begin{pmatrix} 1 & 4 \\ 2 & 5 \\ 3 & 6 \end{pmatrix},$$

which requires three Givens or two Householder transformations. Built up from the three Givens transformations Q is, to five decimals,

$$\mathbf{Q} = \mathbf{G}_3 \cdot \mathbf{G}_2 \cdot \mathbf{G}_1$$

$$= \begin{pmatrix} 1 & 0 & 0 \\ 0 & 0.90582 & 0.42366 \\ 0 & 0.42366 & -0.90582 \end{pmatrix} \begin{pmatrix} 0.26726 & 0.96362 & 0 \\ 0.96362 & -0.26726 & 0 \\ 0 & 0 & 1 \end{pmatrix}$$

$$\cdot \begin{pmatrix} 1 & 0 & 0 \\ 0 & 0.55470 & 0.83205 \\ 0 & 0.83205 & -0.55470 \end{pmatrix}$$

$$= \begin{pmatrix} 0.26726 & 0.53452 & 0.80178 \\ 0.87287 & 0.21822 & -0.43643 \\ 0.40825 & -0.81650 & 0.40825 \end{pmatrix}.$$

$$\mathbf{Q} \cdot \mathbf{A} = \begin{pmatrix} 3.74164 & 8.55232 \\ 0 & 1.96400 \\ 0 & 0 \end{pmatrix}.$$

Built up from the two Householder transformations Q is

$$\mathbf{H}_2 \cdot \mathbf{H}_1 = \begin{pmatrix} 1 & 0 & 0 \\ 0 & -0.14995 & -0.98868 \\ 0 & -0.98869 & 0.14995 \end{pmatrix}$$

$$\cdot \begin{pmatrix} -0.26726 & -0.53452 & -0.80178 \\ -0.53452 & 0.77454 & -0.33818 \\ -0.80178 & -0.33818 & 0.49274 \end{pmatrix}$$

$$= \begin{pmatrix} -0.26726 & -0.53452 & -0.80178 \\ 0.87286 & 0.21821 & -0.43646 \\ 0.40825 & -0.81649 & 0.40824 \end{pmatrix},$$

and

$$\mathbf{Q} \cdot \mathbf{A} = \begin{pmatrix} -3.74164 & -8.55232 \\ -0.00010 & 1.96373 \\ -0.00001 & -0.00001 \end{pmatrix}.$$

To within the errors of round-off and aside from a factor of -1 the two matrices $\mathbf{Q}$ are the same, But $\mathbf{G}_2 \cdot \mathbf{G}_1$ zeroes the elements 2 and 3 of the first column, as does $\mathbf{H}_1$, but

$$\mathbf{G}_2 \cdot \mathbf{G}_1 = \begin{pmatrix} 0.26726 & 0.53452 & 0.80178 \\ 0.96362 & -0.14825 & -0.22237 \\ 0 & 0.83205 & -0.55470 \end{pmatrix},$$

distinct from $\mathbf{H}_1$. Likewise, both $\mathbf{G}_3$ and $\mathbf{H}_2$ zero the element A_{32}, but are distinct even though here $\mathbf{H}_2$ degenerates to a Givens matrix.

There is still another way to form the orthogonal matrix $\mathbf{Q}$, modified Gram–Schmidt (MGS) orthogonalization. This will not be discussed because it offers no advantage over the Householder triangularization. The interested reader is referred to Section 19.2 of Lawson and Hanson's book (1974).

The discussion so far seems to imply that we must apply orthogonal transformations, whether Givens or Householder, to a matrix of size $m \times n$. Sequential accumulation of the equations of condition results in a symmetric matrix of size $n \times n$. The equations of condition, however, may also be sequentially accumulated by use of orthogonal transformations, but at the cost of an increased operation count. The trick consists of reserving space in an array of size less than $m \times n$, say $(n + 1) \times n$, reading in $n + 1$ equations of condition, reducing them to upper triangular form, reading in a new equation of condition at row $n + 1$ of the array, reduce the array to upper triangular, and so forth until the equations of condition are exhausted and we are left with an upper triangular array. Of course, we must also consider the right-hand side and use an array of size $(n + 1) \times (n + 1)$, the $(n + 1)$st column for the right-hand side.

With the Givens transformations we read in one new equation of condition at a time after the first $(n + 1)$ have been reduced. Householder transformations permit reading in blocks of k equations of condition after the first $n + 1$, for which an array of size $(n + k) \times (n + 1)$ is needed, and reducing the resulting system to upper triangular form. The Givens transformation pays no performance penalty for sequential accumulation. The transformation depends on two elements of a column of $\mathbf{A}$, and it is immaterial if one of the elements is already in computer memory or must be read in as part of an equation of condition. The Householder transformation, on the other hand, pays a definite performance penalty for sequential accumulation.

Ignore, for the moment, the right-hand side of the equations of condition. We assign an array of size $(n + k) \times n$ to process k equations of condition at a time. But on the first pass we input $n + k$ equations and reduce them to upper triangular form. After the first pass the $(n + k) \times n$ array contains an $n \times n$ upper triangular matrix, and the lower $k \times n$ part of the array is null. Into the lower part, k equations of condition are read in and the system reduced to upper triangular form. The process is repeated until all of the equations of condition are reduced. This requires q passes, where q is determined from $m = (n + k) + qk$. On the last pass there may not be exactly k equations of conditions left, but this presents no problem; the remaining equations can still be reduced to upper triangular form in the same $(n + k) \times n$ array.

From the operation count derived previously for the Householder transformation, we deduce that the first pass needs $(n + k)n^2 - n^3/3$ operations, as do the subsequent passes for a total of $(n + k)n^2 - n^3/3 + [m - (n + k)]/k[(n + k)n^2 - n^3/3] = (m - n)/k[(n + k)n^2 - n^3/3]$ operations. For small k this is considerably more than the $mn^2 - n^3/3$ operations for nonsequential processing. For example, suppose that $m = 100$ and $n = 5$. Nonsequential processing needs 2,458 operations. If $k = 1$, sequential Householder processing will use up 10,291 operations. As k increases the operation count goes down. Had we selected $k = 5$, the operation count would be 3,958 and for $k = 15$ the count would be 2,902. For $k = 15$, there are seven passes through the reduction algorithm, the last one of which processes only five equations of condition.

Reflecting a bit, however, convinces us that the operation count for sequential processing is too high because of a poorly implemented algorithm. When k is small the $(n + k) \times n$ array is, after the first pass, nearly 50% sparse. The operation count is greatly lowered by taking account of the sparcity. In Figure 5.3 the code should be modified to check if $U(I)$ is of the order of the machine epsilon—in other words $U(I)$ is null—and not perform the operation that involves this element of the Householder transformation. For example, the sentence

$$A(I, K) = A(I, K) - CONS * U(I)$$

would not be executed. By allowing for the sparcity of $(n + k) \times n$, the author has found from actual operation counts on a VAX-11/780 that for k large there is little difference between sequential and nonsequential processing of the equations of condition. For small k, sequential processing involves twice as many operations as nonsequential processing. This is reasonable because for $k = 1$ the Householder transformation degenerates into a Givens transformation, which requires twice as many operations as the Householder transformation.

An important observation is that whether we use the Givens or the Householder transformation and whether we process the equations of condition sequentially or nonsequentially, we need, for large m compared with n, between two and four times as many operations as required by the formation of normal equations and their Cholesky decomposition.

More operations, but perhaps not necessarily more work. The discussion at the end of Section 5.1 shows that forming the normal equations squares the condition number of the matrix of the equations of condition. Some authors use this fact to assert baldly that orthogonal transformations, which leave the condition number of **A** unchanged, handle problems in single-precision that require double-precision when normal equations are used. A favorite example of the nefarious consequences of opting for normal equations is

$$\mathbf{A} = \begin{pmatrix} 1 & 1 \\ \varepsilon & 0 \\ 0 & \varepsilon \end{pmatrix},$$

where ε is of the order of the machine epsilon in single-precision. The normal equations are

$$\mathbf{B} = \mathbf{A}^{\mathrm{T}} \cdot \mathbf{A} = \begin{pmatrix} 1 + \varepsilon^2 & 1 \\ 1 & 1 + \varepsilon^2 \end{pmatrix};$$

B is singular unless we use double-precision.

But the assertion that orthogonal transformations handle problems in single-precision, when normal equations need double-precision, is an over-simplification of the actual state of affairs. We can easily reach such a conclusion by only looking at the condition number, but Mark Twain, in his essay "The Mississippi", shows the danger of drawing conclusions based on one facet of a problem. Because his language is so delightful the relevant passage is quoted here in full, rather than paraphrased.

"In the space of one hundred and seventy-five years the lower Mississippi has shortened itself two hundred and forty-two miles. That is an average of a trifle over one mile and a third per year. Therefore, any calm person, who is not blind or idiotic, can see that in the Old Oolitic Silurian Period, just a million years ago next November, the lower Mississippi River was upward of one million, three hundred thousand miles long, and stuck out over the Gulf of Mexico like a fishing rod. And by the same token any person can see that seven hundred and forty-two years from now the lower Mississippi will be only a mile and three-quarters long, and Cairo and New Orleans will have joined their streets together.... There is something fascinating about science. One gets such wholesale returns of conjecture out of such a trifling investment of fact."

What Mark Twain writes may be somewhat sarcastic, but germaine to the discussion of orthogonal transformations versus normal equations. By concentrating only on the condition number, we overlook other factors. One of these is the number of arithmetic operations. We have seen that orthogonal transformations require between two and four times more operations than normal equations. More operations mean greater statistical accumulation of round-off or truncation error. Depending on the computer used, whether it rounds or truncates results of floating-point operations, and the transformation selected, Givens or Householder, there will be between 1.4 and four times more accumulation of statistical error.

If $m \gg n$ we need double-precision with either normal equations or orthogonal transformations. Suppose that A is $10,000 \times 50$ and $\text{COND}(\mathbf{A}) = 10^3$, by current standards a man-sized, but not horrendous, problem. On most computers single-precision gives about seven or eight decimal digits of representation. Given that $\text{COND}(\mathbf{A}) = 10^3$ and remembering the discussion in Chapter 2, we see that only four or five of those digits remain significant after reducing **A** to upper triangular form. If we opt for the Householder triangularization, the computer performs $2.5 \cdot 10^7$ arithmetic operations. With truncation and its linear build-up of statistical error, all of the significant digits in single-precision are wiped out, regardless of the condition number. If the computer rounds, the build-up is proportional to the square root of the operation count and about four digits are lost. With $COND(\mathbf{A}) = 10^3$, we are assured of, with luck, one significant digit, scarcely a respectable state of affairs. The moral is evident: if $m \gg n$, *both* orthogonal transformations and normal equations require double-precision. The upshot of the discussion is that orthogonal transformations will require between two and four times more labor—genuine labor, not merely operation count—than normal equations, something guaranteed to irritate the payer of the computer bill. Had we been able to get away with single-precision, the increased operation count for orthogonal transformations would be compensated by the single-precision being typically twice as fast as double-precision..

But suppose that $m > n$ but m is not much greater than n. Perhaps now orthogonal transformations can make do with only single-precision? Perhaps yes, but then again perhaps no. Two important papers, one by Golub and Wilkinson (1966) and the other by Jennings and Osborne (1974), both prove that when $m > n$ even orthogonal transformations are plagued by an increased condition number. These papers should be consulted for details, but the essence of the argument is that, as usually happens when ideas from mathematics are implemented on a computer, the finite length of a floating-point number implies that we work not with the ideal matrices **A** and **Q** of this section, but rather with finite approximations to them. The very act of using an approximate matrix can have nefarious consequences. Following an example in Forsythe, Malcolm, and Moler (1977, p. 45), suppose that our matrix **A** has all elements integer except one, that is, 0.1, our computer is binary with $p = 24$, and $COND(\mathbf{A}) = 10^3$. The element 0.1 has no exact representation in binary, which causes an error in the norm of the solution of the order of

$$\frac{\|\Delta X\|}{\|X\|} = 10^3 \cdot 2^{-24} = 6 \cdot 10^{-5}.$$

Just storing the matrix in the computer may bring about an error in the fifth decimal digit of the solution.

The papers by Golub and Wilkinson (1966) and Jennings and Osborne (1974) demonstrate that when a bound for the norm of the absolute error of the solution **X** is derived, a term

$$\varepsilon \ COND(\mathbf{A}^T \cdot \mathbf{A}) \|\mathbf{r}\|_2$$

appears. ε is a small number of the order of the machine epsilon, $\|\mathbf{r}\|_2$ is the Euclidean norm of the vector of the residuals, and $COND(\mathbf{A}^T \cdot \mathbf{A})$ is the squared condition number of the normal equations, even though we do not explicitly form them.

Golub and Wilkinson state emphatically, by using italics: "We conclude that although the use of orthogonal transformations avoids some of the ill effects inherent in the use of the normal equations, the value of $[\mathrm{COND}(\mathbf{A}^{\mathrm{T}} \cdot \mathbf{A})]$ is still relevant to some extent." Jennings and Osborne put the matter this way: "Orthogonal factorization methods for solving an overdetermined system of linear equations, in the least squares sense, are known to possess better stability properties than the method of inverting the normal equations, provided the residual is small.... When the residual is large both methods are dependent on the square of the condition number ... of the matrix." There is, unfortunately, no way to guarantee that $\mathbf{r}$ is small, at least until the data have been filtered. But, data filtering involves having a preliminary solution based on all of the data, to determine an adequate cutoff for what we shall consider discordant data. The preliminary solution will most likely incorporate large residuals.

To summarize the discussion so far, only when $m = n$ can we make the categorical statement that orthogonal transformations need only single-precision, but that normal equations need double-precision. Of course, when $m = n$, it would be stupid to solve the linear system by forming normal equations rather than by a more direct method, such as Gaussian elimination. For $m > n$, we should probably use double-precision with either orthogonal transformations or normal equations, and for $m \gg n$, we definitely should. For $m > n$, but not $m \gg n$, the solution from orthogonal transformations is likely to be better than one from normal equations, but not startlingly so.

To illustrate this last statement consider the system

$$\begin{pmatrix} 1 & 2 \\ 3 & 4 \\ 5 & 10^6 \end{pmatrix} \begin{pmatrix} X_1 \\ X_2 \end{pmatrix} = \begin{pmatrix} 1 \\ 2 \\ 3 \end{pmatrix}. \tag{5.81}$$

The large 10^6 element compared with the rest assures a large residual. The author solved Eq. (5.81) in double-precision, with inner products accumulated in extended precision, on a VAX-11/780 using both Householder transformations and normal equations. The solution from Householder transformation was

$$\begin{pmatrix} X_1 \\ X_2 \end{pmatrix} = \begin{pmatrix} 0.7000006000031200 \cdot 10^0 \\ -0.50000280001400009 \cdot 10^{-7} \end{pmatrix}$$

and from normal equations

$$\begin{pmatrix} X_1 \\ X_2 \end{pmatrix} = \begin{pmatrix} 0.7000006000031200 \cdot 10^0 \\ -5.0000280001400018 \cdot 10^{-7} \end{pmatrix},$$

nearly identical. To three decimals $\mathbf{r}^{\mathrm{T}} = (-0.300, 9.999 \cdot 10^{-2}, 2.000 \cdot 10^{-7})$, a considerable spread in the sizes of the individual residuals.

Then the author solved Eq. (5.81) in single-precision with inner products accumulated in double-precision. The solution from the Householder transformations was

$$\begin{pmatrix} X_1 \\ X_2 \end{pmatrix} = \begin{pmatrix} 0.7000006 \cdot 10^0 \\ -0.50000284 \cdot 10^{-7} \end{pmatrix}$$

and from normal equations

$$\begin{pmatrix} X_1 \\ X_2 \end{pmatrix} = \begin{pmatrix} 0.7000003 \cdot 10^0 \\ -0.50000114 \cdot 10^{-7} \end{pmatrix}.$$

The Householder single-precision solution is undoubtedly better, but the one from the normal equations is surprisingly good. In fact, both involve relative errors in the norm of the solution that are of the order of the machine epsilon in single-precision, just what the discussion in Golub and Wilkinson and Jennings and Osborne would lead us to expect. Accumulating the inner products in a higher precision, either double-precision or extended precision depending on the precision, single or double, with which the rest of the arithmetic operations are carried out, undoubtedly contributes to the closeness of the results.

In general, a least squares reduction must be performed in double-precision whether we use orthogonal transformations or normal equations, and the normal equations involve less work. They also involve less computer memory. Figure 5.1 shows that, by taking advantage of the symmetry of the normal equations, a least squares solution may be carried out in an array of size $n(n + 1)/2$ for the normal equations, an array of size $n + 1$ for each equation of condition, and two ancillary arrays of size n each, a total of $n(n + 1)/2 + 3n + 1$. Orthogonal transformations, on the other hand, require a minimum array size of $(n + 2) \times (n + 1)$—we are now considering the right-hand side also—unless we become entangled with highly convoluted programming to take advantage of the upper triangular characteristic of the reduced system. For large n not only do the normal equations involve less work, they also take up less space.

Orthogonal transformations, therefore, are not exactly the panacea for least squares problems that their proponents sometimes make of them. What, then, is their proper role? That of possibly providing a solution to a poorly conditioned system when normal equations would fail. But first of all we should see what is the source of the poor conditioning.

When forming the matrix **A** of the equations of condition, we should see to it that the columns of **A** do not vary greatly among themselves, which will be evidenced by a high condition number. A poor choice of units, say microns for one variable and light years for another, will cause one column of **A** to be much larger than another. We should either make a more sensible choice of units or precondition **A** so that the variation among the columns is minimized. Mathematically, we postmultiply **A** by a diagonal matrix **S** of scale factors, with the elements of **S** selected to equilibrate the columns of **A**. The elements of **S** could be, for example, the reciprocals of the Euclidean norms of the columns of **A**.

But if the ill-conditioning is inherent to **A** and cannot be improved by preconditioning, the first consideration should be one of obtaining more data to strengthen the system. This book assumes that the reader is a scientist interested in estimating parameters from his data, not a mathematician or numerical analyst. Mathematical technique, no matter how sophisticated, is a poor substitute for more or better data. A poorly conditioned system means that we should leave mathematics aside and start doing science by obtaining more data. A solution eked out by orthogonal transformations, should normal equations fail, in no way compensates for lack of data and is likely to be a dubious contribution to science.

Sometimes, it is true, more data are unobtainable. There is only a finite number of historical records. If our analysis depends critically on these records, we must look to the reduction technique that gives the best results, as there is no way to strengthen the solution by gathering more data. This is the situation where orthogonal transformations become the mathematical tool of choice for the least squares solution of an overdetermined system. Even though they do not possess the virtues their proponents sometimes attribute to them, they may, nevertheless, produce a solution when normal equations fail. Any solution is better than none at all.

But aside from these, generally uncommon, instances, the argument in favor of orthogonal transformations is not as strong as is sometimes made. With sufficient data to form a well-conditioned system and some thought given to preconditioning the matrix of the equations of condition, the normal equations are likely to be entirely adequate, and computationally less costly than orthogonal transformations.

5.5. Iteratively Reweighted Least Squares

Orthogonal transformations may overcome one of the defects of the method of least squares as traditionally practiced, an increase in the condition number of the linear system subsequent to forming the normal equations. Another defect, lack of robustness, is inherent in the criterion of least squares itself, not an ancillary consideration to the method of solution of the system. It is, nevertheless, possible to obtain robustness with the criterion of least squares.

But what, exactly, is robustness? The term was introduced in Section 4.2, to refer to a solution's being little influenced by outliers. Rather than give a precise definition, and even statisticians vary among themselves in defining robustness, a heuristic description may be just as illuminating. A robust procedure is one little influenced by outliers. For example, take the series of numbers 1, 2, 3, 4, 5 and ask what their average is. The usual average is the mean, 3 here. But sometimes the median is used, also 3 for our series. Occasionally, we use the midrange, the mean of the highest and the lowest value in the series, and still 3. Now change the 5 to 10^6. If the numbers had been obtained from an experiment, 10^6 would most likely be a discordant value (but not necessarily; remember the pre-general relativity residuals from Mercury's motion). The mean from the series of numbers jumps to $2 \cdot 10^5$, the midrange shoots up to $5 \cdot 10^5$, but the median remains at 3. Of the three averages, the median is robust because it is little influenced—in this example not influenced at all—by the discordant 10^6. Both the mean and the midrange, the latter more so than the former, are unrobust; the discordant, or supposedly discordant, 10^6 completely ruins the average.

The same considerations apply to the method of least squares in general, of which the mean represents the special case of least squares estimation of the average of a series of numbers. Equation (5.64) has a least squares solution of $X^T = (0.6\ 0.4)$. If the fifth element of the right-hand side is changed from 2 to 100—in practice, a value of 100 compared with the others would most likely be discordant—the solution changes to $X^T = (-38.6\ 20)$, a relative error of 6,000% if we take $X^T = (0.6\ 0.4)$ as the true solution. The solution is decidely unrobust. The L_1 solution of Eq. (5.64), and Chapter 6 will show how the L_1 solution may be obtained, is $X^T = (0.5\ 0.5)$. When the fifth element jumps from 2 to 100 the L_1 solution changes to $X^T = (0\ 1)$, obviously much less influenced by the discordant data point. The L_1 solution is decidely robust.

Iteratively reweighted least squares, discussed by Coleman, Holland, Kaden, Klema, and Peters (1980), are ordinary least squares, the least squares of the normal equations of Section 5.1 or the orthogonal transformation least squares of Section 5.4, applied to a data set whose residuals have been weighted to achieve robustness. To start the process we need a preliminary solution, such as a least squares solution obtained from all of the unfiltered data or, better, an L_1 solution. Let our data model be, once again, Eq. (4.2) of Chapter 4, $r = A \cdot X - d$. In ordinary least squares we minimize $r^T \cdot r$. In iteratively reweighted least squares we minimize

$$\|W^{1/2} \cdot r\|_2 = \|W^{1/2} \cdot A \cdot X - W^{1/2} \cdot d\|_2 = \min, \qquad (5.82)$$

where W is a diagonal matrix of weights. All we have to do is use ordinary least squares with $W^{1/2} \cdot A$ and $W^{1/2} \cdot d$ in place of A and d and keep on iterating until the gradient

$$(W^{1/2} \cdot A)^T \cdot (W^{1/2} \cdot r)$$

is smaller than some tolerance. From Eq. (5.14) an ordinary least squares solution is obtained when the gradient $A^T \cdot r = 0$.

The weights themselves, the diagonal elements of W, may be anything the user considers reasonable. Table 5.1 gives some weights, derived from Table I of Coleman Holland, Kaden, Klema, and Peters (1980), that are frequently used. Table 5.1 leaves out two weigths, Cauchy and Logistic, that Table I of Coleman Holland, Kaden, Klema, and Peters includes. The weights fall into three categories. Andrews, Biweight, and Talwar is merely a recasting of a criterion for outlier rejection, such as Pierce's criterion of Section 4.3. The second category, Welsch, uses no cutoff, but large residuals receive such small weight that they are effectively zero; in fact, given the finite length of floating-

Table 5.1. Weight functions.

Name	Weight	Range
Andrews	$\left. \begin{array}{l} \sin(r/1.339)/(r/1.339) \\ 0 \end{array} \right\}$	$\begin{array}{l} \lvert r \rvert \le 4.207 \\ \lvert r \rvert > 4.207 \end{array}$
Biweight	$\left. \begin{array}{l} [1 - (r/4.685)^2]^2 \\ 0 \end{array} \right\}$	$\begin{array}{l} \lvert r \rvert \le 4.685 \\ \lvert r \rvert > 4.685 \end{array}$
Fair	$1/(1 + \lvert r/1.400 \rvert)$	$\lvert r \rvert < \infty$
Huber	$\left. \begin{array}{l} 1 \\ 1.345/\lvert r \rvert \end{array} \right\}$	$\begin{array}{l} \lvert r \rvert \le 1.345 \\ \lvert r \rvert > 1.345 \end{array}$
Talwar	$\left. \begin{array}{l} 1 \\ 0 \end{array} \right\}$	$\begin{array}{l} \lvert r \rvert \le 2.795 \\ \lvert r \rvert > 2.795 \end{array}$
Welsch	$\exp(-r/2.985)^2$	$\lvert r \rvert < \infty$

r = residual scaled by median of absolute values of nonzero residuals.

point numbers the weight for large residuals will result in underflow and be exactly zero. Fair and Huber, the third category, assign relatively high weight to large residuals. The Fair and Huber weightings embody the idea, mentioned in Section 4.3, that is, as Airy and Glaisher maintained, probability alone cannot assure that an observation should be eliminated outright, only that it is improbable and deserving lower, but not zero, weight. The Fair weighting is about as an extreme expression of such a viewpoint as can be found; even a residual differing by one hundred σ from the mean will receive weight 0.014.

The basic concept of iteratively reweighted least square is easy to understand, but certain practical difficulties should be addressed. To achieve effective weighting, the residuals must be scaled by an appropriate factor. Consider residuals from the observations of the positions of planets, with which the author is familiar. An error of 0.1 radians is impossible. Such an observation would be an obvious blunder. Nevertheless, using the Andrews weight function and setting $r = 0.1$—radians are dimensionless after all—we calculate the high weight of 0.999. But if a typical residual is of the order of 3 seconds of arc, or $1.454 \cdot 10^{-5}$ radians, and we use this as a scale factor, our 0.1 residual becomes 6875, well outside the cutoff and assigned zero weight. Both the median of the absolute values of the nonzero residuals and the mean absolute deviation of the residuals (MAD) (Eq. (4.12) of Chapter 4), are good choices for the scale factor. An alternative, the mean error of unit weight, would be less desirable because of its sensitivity to isolated, large residuals. The paper by Coleman, Holland, Kaden, Klema, and Peters (1980) explains how the constants that appear in the weight functions and the cutoffs are determined.

Other practical difficulties manifest themselves in a concrete example, the solution of Eq. (5.64) with the last 2 on the right-hand side replaced by 100.

We start with the unfiltered least squares solution, already given as $X^T =$ $(-38.6\ \ 20)$. For our weight function we take the Biweight. Both the Andrews and the Biweight weight functions are similar, with Biweight going to zero slightly more rapidly than does Andrews. Of the two, Biweight would generally be preferred for extensive applications, because it involves no subroutine call to a trigonometric function. We work to five decimals. With the given solution the initial residuals and weights are (the residuals are on the left and the weights on the right),

→	−19.6	0.91096
	0.4	0.99996
	19.4	0.91272
	38.4	0.68083
	−38.6	0.67782;

the arrow indicates the median residual. Had we used the MAD the scale factor would be 23.28 instead of 19.6. The mean error of unit weight would give a scale factor of 35.24. The median of the absolute values of the nonzero residuals seems the most appropriate scale factor. The downweighting of high residuals is apparent.

After five iterations we have

	−14.54510	0.92569
	1.20378	0.99948
→	15.95265	0.91096
	30.70153	0.69098
	−50.54960	0.29435

with $X^T = (-29.29398\ \ 15.74888)$ and the gradient equal to $(0.53947\ \ 2.69538)$.

The very next iteration calculates weight zero for the largest residual. A total of seventeen iterations is necessary to reduce the gradient to zero, to five decimals. At this stage, we have

→	− 0.28676	0.91096
	0.41135	0.82130
	0.10946	0.98677
	− 0.19243	0.95939
	−96.49432	0.00000

and $X^T = (0.01513\ \ 0.69811)$. Had the discordant 100 been eliminated at first, the ordinary least squares solution would be $X^T = (0\ \ 0.7)$. The solution using the Biweight weight function has converged to a value close to this, but after a substantial number of iterations. Use of a better initial solution would decrease the iteration count. A start from an L_1 solution, $X^T = (-1\ \ 1)$, converges after twelve iterations. The weight function selected also influences the convergence rate. Twelve iterations would also suffice with Welsch weighting.

Had we elected Talwar weighting and an L_1 start, convergence to $\mathbf{X}^T =$ (0 0.7) is obtained after merely one iteration. Our pleasure at such rapid convergence soon disappears, however, if we start from the unfiltered least squares solution. The program falls into an infinite loop; no residuals are eliminated, and the program keeps assigning unit weight to all residuals and holding the solution at $\mathbf{X}^T = (-38.6\ \ 20)$. Why? Because the scaled largest residual, $-38.6/19.6 = 1.969$, does not exceed the cutoff of 2.795 and is not rejected; nor is it downweighted and assured eventual elimination as it assumes less and less importance in the solution. The same thing would happen were we to apply Pierce's criterion of Section 4.3 to the calculated residuals; none of them would fall under the ban of rejection. The solution would remain at $\mathbf{X}^T = (-38.6\ \ 20)$.

It seems, then, that under certain conditions an iteratively reweighted least squares solution may fail to be robust. Two ways to prevent this are: use an L_1 solution to start the process; or use a weight function other than Talwar with a start from a unfiltered least squares solution. This assures us that large residuals will eventually be eliminated. If for some reason we do not wish to assign higher weight to smaller residuals within the cutoff, as with the Andrews and Biweight weight functions, the first few iterations through the algorithm could be made by using such weighting, before switching to Talwar.

Use of iteratively reweighted least squares requires some thought beforehand, as do all mathematical methods, to assure worthwhile results. The author's somewhat limited experience with the technique (Branham, 1986) leads him to some recommendations. Be prepared to pay a computational price for iteratively reweighted least squares. The simple example given nevertheless required seventeen iterations, with the Biweight weight function, to converge to a gradient that is zero to five decimal places. If we have prior knowledge, such as a previous solution to a similar problem or an instinctive feeling for the cutoff for an acceptable observation, we should use it; convergence will be accelerated. For data that arise from the physical sciences a weight function should be selected that imposes a cutoff for acceptable observations. Whatever merit there may be to arguments, such as Glaisher's and Airy's (see Section 4.3), about our not being justified, from probability alone, in rejecting observations, in practice large residuals almost always correspond to blunders and should be eliminated outright, not assigned small weight. The exact value of the cutoff is somewhat arbitrary.

5.6. Constrained Least Squares

Sometimes the physical situation we are mathematically modelling imposes constraints on the solution. When fitting a curve to experimental data points, for example, we may wish the curve to pass through some of the points, such as the end points. How can we handle this situation?

The simplest way, conceptually, assigns high weight to the constraints to be satisfied, forcing the solution to pass near the corresponding data points. If in

Eq. (5.64) we desire the solution to pass through the data point 3 on the righ-hand side, we could weight the corresponding equation of condition by a factor δ greater than unity,

$$\begin{pmatrix} 1 & 1 \\ 1 & 2 \\ 1 & 3 \\ \delta & 4\delta \\ 1 & 5 \end{pmatrix} \begin{pmatrix} X_1 \\ X_2 \end{pmatrix} = \begin{pmatrix} 1 \\ 1 \\ 2 \\ 3\delta \\ 2 \end{pmatrix}. \tag{5.83}$$

The solution of Eq. (5.83) is

$$\begin{pmatrix} X_1 \\ X_2 \end{pmatrix} = \frac{1}{35 + 15\delta^2} \begin{pmatrix} 25 + 5\delta^2 \\ 10 + 10\delta^2 \end{pmatrix}. \tag{5.84}$$

When $\delta = 1$, Eq. (5.84) reduces to the solution $X^T = (0.60000 \ 0.40000)$ already obtained. But as δ increases Eq. (5.84) increasingly approaches $X^T = (\frac{1}{3} \ \frac{2}{3})$, satisfying more and more closely the constraining equation $X_1 + 4X_2 = 3$. Unfortunately, as δ moves upward so does the condition number of Eq. (5.83). If Eq. (5.83) is formed into normal equations, they have the eigenvalues

$$\lambda_{1,2} = \frac{(43 + 17\delta^2) \pm \sqrt{1709 + 1402\delta^2 + 289\delta^4}}{2}; \tag{5.85}$$

the ratio of the larger eigenvalue, λ_1, to the smaller, λ_2, increases without limit as δ increases.

Although in theory we can make the solution pass as close as we desire to a data point, in practice the condition number limits the achieveable accuracy. A reduction of the system by orthogonal transformations ameliorates the problem, but we still cannot guarantee that the constraint is satisfied exactly (where "exactly" means to within a tolerance close to the machine epsilon). Something more sophisticated is called for, even though we pay the price of a more complicated algorithm, more complicated than merely using a standard least squares algorithm with weights assigned to the equations of condition.

In general terms, the constrained least squares problem may be stated as: minimize the Euclidean norm of

$$A \cdot X = d \tag{5.86a}$$

subject to

$$c \cdot X = b, \tag{5.86b}$$

where A, as in Eq. (4.1), is of dimension $m \times n$, with $m > n$, X is the n vector of the solution, d is an m vector of the data points, c is a matrix of dimension $r \times n$, with $r < n$, and b is an r vector of data points to be satisfied exactly. Notice that in Eq. (4.18) C's dimensions are $n \times n$; in Eq. (5.86b), c could also be of this size, but then it would be pointless to speak of a least squares problem. We force the n vector of the solution to pass through n data points, which determine X exactly; Eq. (5.86a) becomes superfluous. To have a true least squares problem the matrix c should satisfy the condition $r < n$.

Lagrange multipliers are a standard technique for solving systems like Eq. (5.86). We form the Lagrangian function

$$L(X, \lambda) = (A \cdot X - d)^T \cdot (A \cdot X - d) + \lambda(c \cdot X - b), \qquad (5.87)$$

where λ, the Langrange multiplier, imposes the constraining conditions Eq. (5.86b). Without λ, Eq. (5.87) coincides with the Lagrangian function for unconstrained least squares. To minimize the Lagrangian function we differentiate partially with respect to X and λ and set the partial derivatives to zero. Upon rearrangement of terms we obtain

$$A^T \cdot (A \cdot X - d) = 0;$$
$$c \cdot X = b. \qquad (5.88)$$

The first part of Eq. (5.88) specifies the vanishing of the gradient of the first term on the right-hand side of Eq. (5.87) and coincides with Eq. (5.6). Indeed, use of the gradient to derive the normal equations furnishes an alternative to that given by Eqns. (5.1)–(5.6). But the normal equations must be solved subject to the restriction of the second part of Eq. (5.88).

To solve Eq. (5.88), partition the matrix c into two submatrices, c_1 of sizes $r \times r$ and c_2 of size $r \times (n - r)$. Likewise, partition A into A_1 of size $m \times r$ and A_2 of size $m \times (n - r)$. Split X into two components, $X^T = (X_1 + X_2)$, where X_1 is an r vector and X_2 an $(n - r)$ vector. The second part of Eq. (5.88) becomes

$$(c_1 \quad c_2)\binom{X_1}{X_2} = b \qquad (5.89)$$

and may be solved for X_1 in terms of X_2,

$$X_1 = c_1^{-1} \cdot (b - c_2 \cdot X_2). \qquad (5.90)$$

Substitute Eq. (5.90) into the first part of Eq. (5.88), rewritten with partitioned A and X,

$$(A_1 \quad A_2)^T \cdot (A_1 \cdot X_1 + A_2 \cdot X_2 - d) = 0. \qquad (5.91)$$

After a bit of algebra we obtain

$$A'^T \cdot A' \cdot X_2 = A'^T \cdot (d - A_1 \cdot c_1^{-1} \cdot b), \qquad (5.92)$$

where $A' = A_2 - A_1 \cdot c_1^{-1} \cdot c_2$. Solve Eq. (5.92) for X_2 and then find X_1 from Eq. (5.90).

To illustrate the process let us consider the least squares solution of

$$\begin{pmatrix} 1 & 2 & 1 \\ 1 & 2 & 4 \\ 1 & 3 & 9 \\ 1 & 4 & 16 \\ 1 & 5 & 25 \end{pmatrix} \begin{pmatrix} X_1 \\ X_2 \\ X_3 \end{pmatrix} = \begin{pmatrix} 1 \\ 1 \\ 2 \\ 3 \\ 2 \end{pmatrix}, \qquad (5.93)$$

where the solution is constrained to pass through the last two data points. Our problem is then to solve

$$\begin{pmatrix} 1 & 1 & 1 \\ 1 & 2 & 4 \\ 1 & 3 & 9 \end{pmatrix} \begin{pmatrix} X_1 \\ X_2 \\ X_3 \end{pmatrix} = \begin{pmatrix} 1 \\ 1 \\ 2 \end{pmatrix} \qquad (5.94a)$$

subject to

$$\begin{pmatrix} 1 & 4 & 16 \\ 1 & 5 & 25 \end{pmatrix} \begin{pmatrix} X_1 \\ X_2 \\ X_3 \end{pmatrix} = \begin{pmatrix} 3 \\ 2 \end{pmatrix}. \qquad (5.94b)$$

For reference, the unconstrained solution of Eq. (5.93) is $X^T = (-0.40000\ 1.25714\ -0.14286)$.

From Eq. (5.94) we have

$$A_1 = \begin{pmatrix} 1 & 1 \\ 1 & 2 \\ 1 & 3 \end{pmatrix}, \qquad A_2 = \begin{pmatrix} 1 \\ 4 \\ 9 \end{pmatrix}, \qquad c_1 = \begin{pmatrix} 1 & 4 \\ 1 & 5 \end{pmatrix},$$

$$c_2 = \begin{pmatrix} 16 \\ 25 \end{pmatrix}, \qquad b = \begin{pmatrix} 3 \\ 2 \end{pmatrix}, \qquad d = \begin{pmatrix} 1 \\ 1 \\ 2 \end{pmatrix}.$$

c_1 is inverted to give

$$c_1^{-1} = \begin{pmatrix} 5 & -4 \\ -1 & 1 \end{pmatrix}.$$

From Eq. (5.90)

$$\begin{pmatrix} X_1 \\ X_2 \end{pmatrix} = \begin{pmatrix} 7 \\ -1 \end{pmatrix} - \begin{pmatrix} -20 \\ 9 \end{pmatrix} X_3.$$

From Eq. (5.92) we find

$$A' = \begin{pmatrix} 12 \\ 6 \\ 2 \end{pmatrix}$$

and

$$(12\quad 6\quad 2) \begin{pmatrix} 12 \\ 6 \\ 2 \end{pmatrix} X_3 = (12\quad 6\quad 2) \begin{pmatrix} -5 \\ -4 \\ -2 \end{pmatrix},$$

from which

$$184 X_3 = -88,$$

or

$$X_3 = -0.47826,$$

and

$$\begin{pmatrix} X_1 \\ X_2 \end{pmatrix} = \begin{pmatrix} -2.56522 \\ 3.30435 \end{pmatrix}.$$

We easily verify that the solution indeed satisfies the constraints.

To use Eqns. (5.90) and (5.92), the matrix c_1 must be nonsingular. If instead of Eq. (5.94b) our constraining equations were

$$\begin{pmatrix} 1 & 4 & 16 \\ 1 & 4 & 25 \end{pmatrix} \begin{pmatrix} X_1 \\ X_2 \\ X_3 \end{pmatrix} = \begin{pmatrix} 3 \\ 2 \end{pmatrix}$$

the solution would break down immediately because of the singularity of c_1. We should check for this possibility and, if necessary, interchange columns of c and A to assure the nonsingularity of c_1. The interchanges need not be performed explicitly. By use of an auxiliary one-dimensional array, such as the NEXT array of Section 3.3.2, to hold index positions of the columns, the interchanges may be done implicitly, with no actual data movement.

But checking for linear dependence of the columns is itself rather time-consuming. Techniques for the purpose, such as calculating the eigenvalues or the determinant of c_1, require of the order in n^3 operations. To be specific, the calculation of the eigenvalues needs about $4n^3/3$ operations and of the determinant $n^3/3$. The advantage lies with calculating the determinant. Figure 5.4 shows a FORTRAN subroutine to calculate the determinant of a square matrix. The code triangularizes the matrix, incorporating the pivot searches discussed in Section 4.4, and calculates the determinant as the product of the diagonal elements. The concepts involved are the same as those presented in Section 4.4 for the solution of a linear system by triangular factorization. If the determinant of c_1 is small, of the order of the machine epsilon, we should perform the column interchanges to produce a nonsingular c_1.

The matrix c_1 may be inverted by Eq. (5.50). Let c_T be the upper triangular matrix that results from the application of an orthogonal transformation Q to c_1. Then

$$c_1^{-1} = c_T^{-1} \cdot Q. \tag{5.95}$$

Because c_T is triangular its inverse is easily calculated by formulas already given in this chapter.

Rather than bother with c_1^{-1}, Lawson and Hanson (1974, Chap. 21) recommend a reduction of Eq. (5.91) by orthogonal transformations that obviate the need for explicitly calculating c_1^{-1}.

This section has been concerned with the method of least squares subject to equality constraints. Inequality constraints are also possible, but the theory becomes rather involved and, because inequality constraints are far less common in practice than equality constraints, will not be presented here. See Lawson and Hanson (1974, Chap. 23) for a discussion. Much of the terminology with inequality constraints comes from linear programming, discussed in the next chapter.

To be complete, a discussion of linear least squares should include the problem of total least squares, where we consider that the matrix A of the equations of condition also incorporates errors, not just the vector d of observed data. This important topic will be deferred until Chapter 8, because the implementation of total least squares depends on the singular value decomposition (SVD), presented in that chapter.

```
C
C
C      THIS PROGRAM FORMS AND SOLVES NORMAL EQUATIONS BY CHOLESKY DECOMPOSITION USING
C      SYMMETRIC STORAGE MODE. IT USES AN ARRAY OF SIZE N * ( N + 1 ) / 2 FOR THE NORMAL
C      EQUATIONS , TWO AUXILIARY ARRAYS OF SIZE  N  FOR THE VECTORS  X  AND  Y , AND AN ARRAY
C      OF SIZE N + 1 FOR THE I - TH  ROW OF THE MATRIX OF THE EQUATIONS OF CONDITION. THE
C      ARRAY  A  IS REPLACED BY THE INVERSE OF THE CHOLESKY FACTOR  S. A SYSTEM OF UP TO
C      SIZE 100 x 100 MAY BE SOLVED.
C
C
                  PROGRAM  NORSLVE
                  IMPLICIT REAL * 8  ( A - H , O - Z )
                  REAL * 16  SUM
                  CHARACTER * 40  FILENAME
                  DIMENSION  A ( 5050 ) , Y ( 100 ) , X ( 100 ) , CONDEQ ( 101 )
                  TYPE * , ' HOW MANY UNKNOWNS ARE THERE ?  '
                  ACCEPT * , N
                  TYPE * , ' WHAT IS THE NAME OF THE DATA-FILE ?  '
                  READ  ( * , 5 )  FILENAME
5                 FORMAT  ( A40 )
                  OPEN  ( 1 , FILE = FILENAME , STATUS = ' OLD ' )
C
C      INITIALIZE .
C
                  DO  I = 1 , N
                    DO  J = I , N
                      A ( J * ( J - 1 ) / 2 + I ) = 0.0
                    END  DO
                    Y ( I ) = 0.0
                    X ( I ) = 0.0
                  END  DO
C
C      INPUT EQUATIONS OF CONDITION AND FORM NORMAL EQUATIONS IN ARRAY  A .
C
                  M = 0
1                 READ  ( 1 , 10 , END = 15 )  ( CONDEQ ( I ) , I = 1 , N + 1 )
10                FORMAT  ( D22.15 )
                  M = M + 1
                  RHS = CONDEQ ( N + 1 )
                  DO  I = 1 , N
                    DO  J = I , N
                      A ( J * ( J - 1 ) / 2 + I ) = A ( J * ( J - 1 ) / 2 + I ) + CONDEQ ( I ) * CONDEQ ( J )
                    END  DO
                    Y ( I ) = Y ( I ) + CONDEQ ( I ) * RHS
                  END  DO
                  GOTO  1
15                TYPE * , ' THERE ARE ' , M , ' EQUATIONS OF CONDITION '
                  IF  ( M . LT . N )  STOP ' SYSTEM IS UNDERDETERMINED '
C
C      DECOMPOSE THE NORMAL EQUATIONS . A  IS REPLACED BY CHOLESKY FACTOR  S .
C
```

Figure 5.1. Solution of normal equations with compact storage mode.

```
            A(1) = SQRT(A(1))
            DO J = 2,N
              A(J * (J-1)/2 + 1) = A(J * (J-1)/2 + 1)/A(1)
            END DO
            DO I = 2,N
              SUM = 0.0
              DO K = 1,I-1
                SUM = SUM + A(I * (I-1)/2 + K) ** 2
              END DO
              A(I * (I-1)/2 + I) = SQRT(A(I * (I-1)/2 + I)-SUM)
              DO J = I + 1,N
                SUM = 0.0
                DO K = 1,I-1
                  SUM = SUM + A(I * (I-1)/2 + K) * A(J * (J-1)/2 + K)
                END DO
                A(J * (J-1)/2 + I) = (A(J * (J-1)/2 + I)-SUM)/A(I * (I-1)/2 + I)
              END DO
            END DO
C
C  FIRST STAGE OF SOLUTION OF NORMAL EQUATIONS. STRANS * Y = D. SOLVE FOR Y.
C
            Y(1) = Y(1)/A(1)
            DO I = 2,N
              SUM = 0.0
              DO K = 1,I-1
                SUM = SUM + A(I * (I-1)/2 + K) * Y(K)
              END DO
              Y(I) = (Y(I)-SUM)/A(I * (I-1)/2 + I)
            END DO
C
C  SECOND STAGE OF SOLUTION OF NORMAL EQUATIONS. S * X = Y. SOLVE FOR X.
C
            X(N) = Y(N)/A(N * (N-1)/2 + N)
            DO I = N-1,1,-1
              SUM = 0.0
              DO K = I + 1,N
                SUM = SUM + A(K * (K-1)/2 + I) * X(K)
              END DO
              X(I) = (Y(I)-SUM)/A(I * (I-1)/2 + I)
            END DO
            TYPE * , ' THE SOLUTION VECTOR  ',(X(I),I = 1,N)
            TYPE * , '
C
C  RE- READ THE EQUATIONS OF CONDITION TO CALCULATE THE RESIDUALS AND SIGMA.
C
            IF (M.GT.N) THEN
              REWIND 1
              SIGMA = 0.0
              TYPE * , '
              TYPE * , ' THE RESIDUALS:  '
```

Figure 5.1 (*continued*)

```
          DO I = 1, M
            READ (1, 10) (CONDEQ(J), J = 1, N + 1)
            SUM = 0.0
            DO J = 1, N
              SUM = SUM + CONDEQ(J) * X(J)
            END DO
            RESID = SUM - CONDEQ(N + 1)
            SIGMA = SIGMA + RESID ** 2
            TYPE * , I, RESID
          END DO
          SIGMA = SQRT ( SIGMA / DFLOAT ( M - N ))
          TYPE * , '
          TYPE * , ' THE MEAN ERROR OF UNIT WEIGHT IS : ', SIGMA
          TYPE * , '
        END IF
C
C    CALL SUBROUTINE TO INVERT THE CHOLESKY FACTOR S. THEN CALCULATE THE COVARIANCE
C    MATRIX.
C
          CALL TRI - INV ( A , N )
          TYPE * , ' THE COVARIANCE MATRIX : '
          DO I = 1, N
            DO J = I, N
              SUM = 0.0
              DO K = J, N
                SUM = SUM + A( K * ( K - 1 )/2 + I ) * A( K * ( K - 1 )/2 + J )
              END DO
              A( J * ( J - 1 )/2 + I ) = SUM
              TYPE * , I, J, A( J * ( J - 1 )/2 + I )
            END DO
          END DO
C
C    CALCULATE THE ESTIMATED ERRORS OF THE SOLUTION X.
C
          IF ( M . GT . N ) THEN
            TYPE * , '
            TYPE * , ' THE ERRORS OF THE SOLUTION VECTOR : '
            DO I = 1, N
              ERR = SIGMA * SQRT( A( I * ( I - 1 )/2 + I ))
              TYPE * , I, ERR
            END DO
          END IF
C
C    CALCULATE THE CORRELATION MATRIX.
C
          TYPE * , '
          TYPE * , ' THE CORRELATION MATRIX : '
          DO I = 1, N
            DO J = I, N
              IF ( I . EQ . J ) THEN
                CORR = 1.0
```

Figure 5.1 (*continued*)

```
              ELSE
                CORR = A ( J * ( J - 1 ) / 2 + I ) /
1                ( SQRT ( A ( I * ( I - 1 ) / 2 + I ) ) * SQRT ( A ( J * ( J - 1 ) / 2 + J ) ) )
              END  IF
              TYPE * , I , J , CORR
            END  DO
          END  DO
        END
C
C     SUBROUTINE FOR IN - PLACE INVERSION OF A TRIANGULAR MATRIX
C
          SUBROUTINE TRI – INV ( A , N )
          IMPLICIT REAL * 8  ( A - H , O - Z )
          REAL * 16  SUM
          DIMENSION  A ( 5050 )
C
C     CALCULATE INVERSE OF DIAGONAL ELEMENTS .
C
          DO  I = 1 , N
            A ( I * ( I - 1 ) / 2 + I ) = 1.0 / A ( I * ( I - 1 ) / 2 + I )
          END  DO
C
C     CALCULATE INVERSE OF OFF - DIAGONAL ELEMENTS .
C
          DO  I = 1 , N - 1
            DO  J = I + 1 , N
              SUM = 0.0
              DO  L = I , J - 1
                SUM = SUM + A ( L * ( L - 1 ) / 2 + I ) * A ( J * ( J - 1 ) / 2 + L )
              END  DO
              A ( J * ( J - 1 ) / 2 + I ) = - A ( J * ( J - 1 ) / 2 + J ) * SUM
            END  DO
          END  DO
          RETURN
          END
```

Figure 5.1 (*continued*)

```
C
C
C      PROGRAM TO IMPLEMENT TRIANGULARIZATION OF AN M x N MATRIX USING GIVENS
C      ROTATION MATRICES .
C
C

       PROGRAM GIVENS
       IMPLICIT REAL * 8 ( A - H , O - Z )
       DIMENSION  A ( 100 , 100 )
C
C      INPUT THE MATRIX A
C
       TYPE * , 'WHAT IS M ? '
       ACCEPT * , M
       TYPE * , 'WHAT IS N ? '
       ACCEPT * , N
       DO I = 1 , M
         DO J = 1 , N
           TYPE * , 'A ( ', I, ', ', J, ') = ? '
           ACCEPT * , A ( I , J )
         END DO
       END DO
C
C      BEGIN THE TRANSFORMATIONS .
C
       DO J = 1 , N
         DO I = M , J + 1 , - 1
C
C      CONSTRUCT THE ELEMENTS  S  AND  C  OF THE GIVENS ROTATION MATRIX .
C
           T = MAX ( ABS ( A ( I , J ) ) , ABS ( A ( I - 1 , J ) ) )
           U = MIN ( ABS ( A ( I , J ) ) , ABS ( A ( I - 1 , J ) ) )
           IF ( T . NE . 0.0 ) THEN
             DENOM = T * SQRT ( 1 D 0 + ( U / T ) ** 2 )
           ELSE
               DENOM = 0.0
           END IF
           IF ( DENOM . NE . 0.0 ) THEN
             C = A ( I - 1 , J ) / DENOM
             S = A ( I , J ) / DENOM
           ELSE
             C = 1.0
             S = 0.0
           END IF
C
C      APPLY THE TRANSFORMATION TO THE MATRIX ELEMENTS .
C
           DO K = J , N
             T1 = C * A ( I - 1 , K ) + S * A ( I , K )
             IF ( K . EQ . J ) THEN
                 T2 = 0.0
             ELSE
                 T2 = - S * A ( I - 1 , K ) + C * A ( I , K )
             END IF
             A ( I - 1 , K ) = T1
             A ( I , K ) = T2
           END DO
         END DO
       END DO
C
C      PRINT OUT THE TRIANGULARIZED MATRIX .
C
       WRITE ( * , 5 ) ( ( I , J , A ( I , J ) , J = 1 , N ) , I = 1 , M )
 5     FORMAT ( 2 ( I2 , '  ' ) , D22 . 15 )
       END
```

Figure 5.2. Givens triangularization of $m \times n$ matrix.

```
C
C
C     TRIANGULARIZE AN  M x N  MATRIX  A  AND THE RIGHT-HAND-SIDE USING HOUSEHOLDER
C     TRANSFORMATIONS.
C
C
      PROGRAM HOUSEHOLDER
      IMPLICIT REAL * 8  (A - H , O - Z)
      DIMENSION  A (100, 100) , U (100) , X (100)
C
C     INPUT THE MATRIX A
C
      TYPE * , 'WHAT IS M ? '
      ACCEPT * , M
      TYPE * , 'WHAT IS N ? '
      ACCEPT * , N
      DO  I = 1 , M
         DO  J = 1 , N
            TYPE * , 'A ( ', I, ', ', J, ') = ? '
            ACCEPT * , A ( I , J )
         END DO
      END DO
C
C     INPUT THE RIGHT-HAND -SIDE , STORED IN COLUMN A ( I , N + 1 ) OF MATRIX A .
C
      DO I = 1 , M
         TYPE * , 'B ( ', I, ') = ? '
         ACCEPT * , A ( I , N + 1 )
      END DO
C
C     BEGIN THE TRANSFORMATIONS .
C
      DO J = 1 , N + 1
         DO I = 1 , M
            U ( I ) = A ( I , J )
         END DO
         CALL  H ( M , J , U , DIVSOR )
C
C     APPLY THE TRANSFORMATION TO THE COLUMNS OF A .
C
         DO K = J , N + 1
            SUM = 0.0
            DO I = J , M
               SUM = SUM + U ( I ) * A ( I , K )
            END DO
            CONS = SUM / DIVSOR
            DO I = J , M
               A ( I , K ) = A ( I , K ) - CONS * U ( I )
            END DO
         END DO
      END DO
C
C     PRINT OUT THE TRIANGULARIZED MATRIX .
C
```

Figure 5.3. Householder triangularization of $m \times n$ matrix and right-hand side.

```
          WRITE ( * , 5 ) ( ( I , J , A ( I , J ) , J = 1 , N + 1 ) , I = 1 , M )
5         FORMAT ( 2 ( I 2 , '  ' ) , D 22.15 )
          END
C
C   SUBROUTINE TO CONSTRUCT THE VECTOR  U  AND THE CONSTANT  2 * MU  NEEDED IN
C   THE HOUSEHOLDER TRANSFORMATION .
C
          SUBROUTINE  H ( M , L , U , DIVSOR )
          IMPLICIT REAL * 8 ( A - H , O - Z )
          REAL * 8 MAX
          REAL * 16 SUM
          DIMENSION U ( 1 )
          MAX = 0.0
          DO I = L , M
              IF ( ABS ( U ( I ) ) . GT . MAX ) MAX = ABS ( U ( I ) )
          END DO
          SUM = 0.0
          DO I = L , M
              SUM = SUM + ( U ( I ) / MAX ) * * 2
          END DO
          SUM = MAX * * 2 * SUM
          S = SQRT ( SUM )
          IF ( U ( L ) . LT . 0.0 ) S = - S
          DIVSOR = SUM + U ( L ) * S
          U ( L ) = U ( L ) + S
          RETURN
          END
```

Figure 5.3 (*continued*)

```
C
C
C   SUBROUTINE TO EVALUATE THE DETERMINANT OF AN N x N MATRIX . DETERMINANT
C   CALCULATED AS PRODUCT OF DIAGONAL ELEMENTS OF THE TRIANGULARIZED MATRIX .
C
C

          SUBROUTINE  DETERM ( AA , N , D )
          IMPLICIT  REAL * 8  ( A - H , O - Z )
          DIMENSION   AA ( 100 , 100 ) , A ( 100 , 100 )
          COMMON  / A /  EPS
C
C   PRESERVE COPY OF ORIGINAL MATRIX .
C
          DO J = 1 , N
            DO I = 1 , N
              A ( I , J ) = AA ( I , J )
            END DO
          END  DO
          D = 1D0
          K = 1
C
C   K  IS BOOKKEEPING INDEX . START WITH  K = 1  AND GO UP TO  K = N .
C
          DO  WHILE  ( K . NE . N )
            IR = K
            IC = K
            B = ABS ( A ( K , K ) )
C
```

Figure 5.4. Calculation of determinant of $n \times n$ matrix.

```
C    SEARCH FOR LARGEST ELEMENT OF ARRAY IN ABSOLUTE VALUE .
C
              DO I = K , N
                DO J = K , N
                  IF ( ABS ( A ( I , J ) ) . GT . B )  THEN
                    IR = I
                    IC = J
                    B = ABS ( A ( I , J ) )
                  END IF
                END DO
              END DO
C
C    INTERCHANGE ROWS .
C
              IF ( IR . GT . K )  THEN
                DO J = K , N
                  C = A ( IR , J )
                  A ( IR , J ) = A ( K , J )
                  A ( K , J ) = -C
                END DO
              END IF
C
C    INTERCHANGE COLUMNS .
C
              IF ( IC . GT . K )  THEN
                DO I = K , N
                  C = A ( I , IC )
                  A ( I , IC ) = A ( I , K )
                  A ( I , K ) = -C
                END DO
              END IF
C
C    DIAGONAL ELEMENT OF ORDER OF MACHINE EPSILON INDICATES SINGULAR MATRIX .
C
              IF ( ABS ( A ( K , K ) ) . LT . EPS )  THEN
                D = 0D0
                RETURN
              ELSE
                D = A ( K , K ) * D
              END IF
C
C    TRIANGULARIZE ARRAY .
C
              DO J = K + 1 , N
                A ( K , J ) = A ( K , J ) / A ( K , K )
                DO I = K + 1 , N
                  B = A ( I , K ) * A ( K , J )
                  A ( I , J ) = A ( I , J ) -B
                END DO
              END DO
              K = K + 1
              END DO
C
C    CALCULATE FINAL VALUE FOR DETERMINANT .
C
              D = A ( N , N ) * D
              RETURN
              END
```

Figure 5.4 (*continued*)

References

Branham, R.L., Jr. (1986). Is Robust Estimation Useful for Astronomical Data Reduction?, *Quarterly J. Royal Astron. Soc.*, **27**, p. 182.

Coleman, D., Holland, P., Kaden, N., Klema, V., and Peters, S.C. (1980). A System of Subroutines for Iteratively Reweighted Least Squares Computations, *ACM Trans Math. Software*, **6**, p. 327.

Faddeeva, V.N. (1959). *Computational Methods of Linear Algebra* (Dover, New York).

Forsythe, G., Malcolm, M.A., and Moler, C.B. (1977). *Computer Methods for Mathematical Computations* (Prentice-Hall, Englewood Cliffs, N.J.).

Forsythe, G. and Moler, C.B. (1967). *Computer Solution of Linear Algebraic Systems* (Prentice-Hall, Englewood Cliffs, N.J.).

Givens, W. (1954). Numerical Computation of the Characteristic Values of a Real Symmetric Matrix, Oak Ridge Nat. Lab. Report ORNL—1574 (Oak Ridge, Tenn.).

Golub, G.H. and Wilkinson, J.H. (1966). Note on the Iterative Refinement of Least Squares Solution, *Numer. Math.*, **9**, p. 139.

Hanson, R.J. (1973). Is the Fast Givens Transformation Really Fast?, *ACM SIGNUM Newsletter*, **8**, p. 7.

Householder, A.S. (1958). Unitary Triangularization of a Nonsymmetric Matrix, *J. ACM.* **5**, p. 339.

Jennings, L.S. and Osborne, M.R. (1974). A Direct Error Analysis for Least Squares, *Numer. Math.*, **22**, p. 325.

Lawson, C.L. and Hanson, R.J. (1974). *Solving Least Squares Problems* (Prentice-Hall, Englewood Cliffs, N.J.).

Wilkinson, J.H. (1965). *The Algebraic Eigenvalue Problem* (Oxford University Press, Oxford).

CHAPTER 6

The L_1 Method

6.1. Introduction

In Chapter 4 we pointed out that the L_1 criterion, i.e., minimize the sum of the absolute values of the residuals

$$\|\mathbf{r}\|_1 = \|\mathbf{A} \cdot \mathbf{X} - d\|_1 = \min, \tag{6.1}$$

is, because of its robustness, extremely useful for data reduction. Where does this robustness come from? The answer to this question will be given in several stages.

To begin with, a large residual enters with only linear weight with the L_1 criterion,

$$\|\mathbf{r}\|_1 = \sum_{i=1}^{m} |r_i|, \tag{6.2}$$

but with quadratic weight, not completely ameliorated by taking the square root, with the least squares criterion,

$$\|\mathbf{r}\|_2 = \left(\sum_{i=1}^{m} r_i^2 \right)^{1/2}. \tag{6.3}$$

If, for example, our vector of residuals were $\mathbf{r}^T = (0.1 \ \ 0.2 \ \ -0.15 \ \ 5 \ \ -0.25)$, the 5 is most likely discordant. But this one value swamps the others when we sum the squares of the residuals, giving an unrealistic mean error of unit weight, $\sigma(1)$. The mean absolute deviation (MAD) (see Eq. (4.12)), although still influenced by the discordant 5, reflects it to a lesser extent: $\sigma(1) = 2.507$; MAD $= 1.425$.

An L_1 solution corresponds to a distribution of error of the form

$$f(X) = \tfrac{1}{2} h e^{-h|X|}, \tag{6.4}$$

where h is called the modulus of precision. Like σ of the normal distribution (see Eq. (4.9)), h measures the concentration of the distribution: large h means a narrow distribution. (For the normal distribution $\sigma = 1/\sqrt{2}h$; for the L_1 distribution there is no σ, and we use h directly.) A large h implies a narrow distribution because the h in the exponential dominates the linear h.

We may easily show that the frequency distribution of error of Eq. (6.2) corresponds with the L_1 criterion. Paralleling the development of Eqns. (4.13)–(4.17), we have for our probability distribution

$$p(d_i, \mathbf{X}) = \tfrac{1}{2}he^{-h|r_i|} = \tfrac{1}{2}h \exp\left[-h\left|d_i - \sum_{j=1}^{n} a_{ij}X_j\right|\right]. \tag{6.5}$$

For the likelihood function we obtain, because once again we assume that the d_i are independent,

$$L(\mathbf{X}) = \prod_{i=1}^{m} \tfrac{1}{2}he^{-h|r_i|} = (\tfrac{1}{2}h)^m \exp\left[-\sum_{i=1}^{m} |r_i|\right]. \tag{6.6}$$

The likelihood function reaches it maximum, satisfying Eq. (4.15), when the exponent of Eq. (6.4) is a minimum,

$$\sum_{i=1}^{m} |r_i| = \min, \tag{6.7}$$

the L_1 criterion.

A comparison of Eq. (6.4) with the normal curve, Eq. (4.9), helps demonstrate why the L_1 criterion is more robust than the least squares criterion: Eq. (6.4) goes to zero far more slowly than Eq. (4.9) and, therefore, encompasses, as normal, large residuals that would be extremely improbable with the least squares criterion. To be more specific, let us consider the probability of encountering a residual equal to or exceeding a certain size with both the L_1 and the least squares criteria, as shown in Table 6.1. This table was prepared from integration of Eqns. (6.4) and (4.9) between the limits $-k\sigma$ and $k\sigma$, where k is a constant multiplier, and keeping in mind that the h of Eq. (6.4) corresponds to $1/\sqrt{2}\sigma$ of Eq. (4.9). If the limits are $-\infty$ and ∞ both of Eqns. (6.4) and (4.9) yield an integral of unity, as they should because the probability of encountering a residual of any size between infinite limits is unity. But for other limits we subtract the integral from unity to find the probability of having a residual exceeding the limit. Equation (6.4) may be integrated easily enough, but Eq. (4.9) presents greater difficulties. Zelen and

Table 6.1. Probability of the residual exceeding the limit.

$k\sigma$	$1 - \int_{-k\sigma}^{k\sigma} f(X), L_1$	$1 - \int_{-k\sigma}^{k\sigma} f(X)$, least squares
1	0.493	0.317
2	0.243	$0.455 \cdot 10^{-1}$
3	0.120	$0.270 \cdot 10^{-2}$
5	$0.291 \cdot 10^{-1}$	$0.573 \cdot 10^{-6}$
10	$0.849 \cdot 10^{-3}$	$0.152 \cdot 10^{-22}$
20	$0.721 \cdot 10^{-6}$	$0.551 \cdot 10^{-88}$

Severo (1965) provide tables and formulas for integrating Eq. (4.9) between the established limits. Without going into details, which are irrelevant here, we can state that Eq. (4.9) cannot be evaluated in closed from. Use of the power series expansion for the exponential permits a term-by-term evaluation of the integral, convergent for all values of X but inefficient for large X; if X is large an asymptotic expansion is more efficient.

A ten σ error is so improbable with the least squares criterion that a residual of this size would never be encountered by chance fluctuations alone. The L_1 criterion, on the other hand, considers that such a residual will be found almost once in a thousand occurrences. Even a twenty σ residual, impossible with least squares, may be expected with the L_1 criterion once in a data set of size one million, a large data set but not inconceivably large. With this consideration in mind we understand that a large residual is far less pernicious with the L_1 criterion than with least squares.

A further explanation of the L_1 criterion's robustness ensues from considering the problem of fitting a straight line to the three discordant data points of Figure 6.1. The L_1 criterion passes the line through the two end points; only point 2 has a nonzero residual. Why?

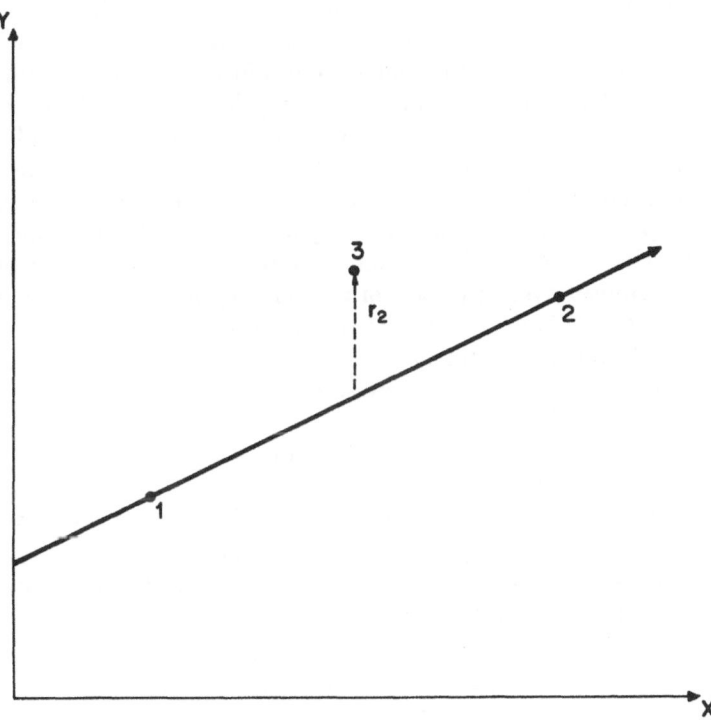

Figure 6.1. Fitting straight line to three discordant data points.

We affirm that

$$\sum_{i=1}^{3} |r_i| = |r_2| = \text{min.} \tag{6.8}$$

Suppose that we displace the line parallel to itself towards point 2 by an amount ε. Instead of (6.8) we have

$$\sum_{i=1}^{3} |r_i| = |\varepsilon| + |r_2 - \varepsilon| + |\varepsilon|. \tag{6.9}$$

ε is positive, and by the relation

$$|X - y| \geq |X| - |y|. \tag{6.10}$$

Equation (6.9) becomes

$$\sum_{i=1}^{3} |r_i| = 2\varepsilon + |r_2 - \varepsilon| \geq \varepsilon + r_2, \tag{6.11}$$

proving the affirmation (6.8): any line that does not pass through two of the points increases the sum of the absolute values of the residuals.

No matter what the size of r_2 may be, the L_1 criterion passes the solution through points 1 and 3, point 2 having little say in the matter. (Little say, but some say. If r_2 becomes too large the L_1 criterion, satisfying (6.4), passes the solution through points 2 and 3 or 1 and 2, point 1 or point 3 becoming the one which nonzero residual). Should r_2 be large, most likely corresponding to discordant data, it has no more pull on the solution than a concordant data point with small r_2. The least squares straight line, on the other hand, pays great attention to point 2. The straight line passes through none of the three points, but rather through what might be called their center of mass. If all three points are nearly collinear, the line passes close to all three, but a discordant point 2 pulls the solution away from the good data points 1 and 3.

Fitting a straight line to three points exemplifies the general case of solving an overdetermined system of size $m \times n$ ($m \geq n$): n of the m equations are satisfied exactly, giving rise to n zero residuals. This assertion will be proved in Section 6.3. In the meantime the following argument makes it reasonable. With m residuals, n of which are zero, we have

$$\sum_{i=1}^{m-n} |r_i| = \text{min.} \tag{6.12}$$

Suppose that (6.12) is not true, and that by increasing the zero residuals by small amounts and decreasing n of the $m - n$ nonzero residuals by the same small amounts, we hope to decrease the sum of the absolute values of the residuals. That is, we affirm that

$$\sum_{i=1}^{m-2n} |r_i| + \sum_{j=m-2n+1}^{m-n} |r_j - \varepsilon_j| + \sum_{k=1}^{n} |\varepsilon_k| \tag{6.13}$$

is smaller than (6.12). The ε_j's and the ε_k's are the same. By the inequality (6.10) we obtain

$$\sum_{i=1}^{m-2n} |r_i| + \sum_{j=m-2n+1}^{m-n} |r_j - \varepsilon_j| + \sum_{k=1}^{n} |\varepsilon_k| \geq \sum_{i=1}^{m-2n} |r_i|$$

$$+ \sum_{j=m-2n+1}^{m-n} |r_j| - \sum_{j=m-2n+1}^{m-n} |\varepsilon_j| + \sum_{k=1}^{n} |\varepsilon_k| = \sum_{i=1}^{m-n} |r_i|.$$

The terms in the ε's cancel because j and k are dummy indices. At best, relation (6.13) is equal to Eq. (6.12), but may be greater. In general, having n of the residuals zero minimizes the sum of the absolute values of the residuals.

That n of the m equations of the matrix A are satisfied exactly and the remaining equations, in a certain sense, ignored—point 2 of Figure 6.1 only influences the solution when r_2 becomes so large that points 2 and 3 or 1 and 2 are satisfied and point 1 or point 3 becomes discordant—is philosophically objectionable to some. Gauss expressed some uneasiness about this matter when he wrote in his *Teoria Motus* (1809, p. 270):

It can be easily shown, that a system of values of unknown quantites, derived from this principal alone [the L_1 criterion], must necessarily exactly satisfy as many equations out of the number proposed, as there are unknown quantities, so that the remaining equations come under consideration only so far as they help to *determine the choice*: if, therefore, the equation $V = M$, for example, is of the number of those which are not satisfied. the system of values found according to this principle would in no respect be changed, even if any other value N had been observed instead of M ....

Despite these philosophical objections—or prejudices—the L_1 criterion is used when robustness is needed. As we have seen an L_1 solution is exceedingly robust.

6.2. General Considerations on the L_1 Solution

How do we obtain an L_1 solution to the overdetermined system of Eq. (4.1)? The fact that exactly n of the m equations are satisfied provides a simple way: solve all of the combinations of equations and select the solution that minimizes the sum of the absolute values of the residuals. Equation (6.14), taken from Barrodale and Robert's paper on an L_1 algorithm (1973),

$$\begin{pmatrix} 1 & 1 \\ 1 & 2 \\ 1 & 3 \\ 1 & 4 \\ 1 & 5 \end{pmatrix} \begin{pmatrix} X_1 \\ X_2 \end{pmatrix} = \begin{pmatrix} 1 \\ 1 \\ 2 \\ 3 \\ 2 \end{pmatrix} \tag{6.14}$$

may be solved for, in the L_1 norm, by taking the first row, combining it with the second, third, and so-forth row, obtaining a solution, calculating the sum of the absolute values of the residuals; then starting with the second row we do the same with the third and following rows; and so forth. Table 6.2 shows the results.

Table 6.2. Solutions of Eq. (6.12).

First row	Second row	X_1	X_2	$\sum \lvert r_i \rvert$
1	2	1	0	4
1	3	0.5	0.5	2
1	4	$\frac{1}{3}$	$\frac{2}{3}$	3
1	5	0.75	0.25	2
2	3	-1	1	3
2	4	-1	1	3
2	5	$\frac{1}{3}$	$\frac{1}{3}$	$2\frac{1}{3}$
3	4	-1	1	3
3	5	2	0	3
4	5	7	-1	11

Two observations ensue from an examination of Table 6.2: not all of the solutions are independent; rows 2 and 3, 2 and 4, and 3 and 4 have the same solution of $\mathbf{X}^T = (-1 \ 1)$; and two independent solutions, $\mathbf{X}^T = (0.75 \ 0.25)$ and $\mathbf{X}^T = (0.5 \ 0.5)$, have the same minimum sum of the absolute values of the residuals. In a simple example like this—with all of the numbers integers and small ones too—that not all of the solutions are independent should not surprise us too much. But two independent solutions that have the same sum of the absolute values of the residuals appears, at first sight, rather startling. This is an instance of a degenerate solution. The next section explains why it may occur. With large matrices **A** and data taken from experiments or observations, the chance of encountering a degenerate solution is slight. But should one occur either solution is acceptable. Here we would probably select $\mathbf{X}^T = (0.5 \ 0.5)$ because it is closer to the least squares solution of $\mathbf{X}^T = (0.6 \ 0.4)$.

This simple method for obtaining an L_1 solution is, however, completely impractical for other than trivially small systems. There is a total of $m!/(m - n)! \, n!$ combinations of equations. Even for moderately sized m and n the factorials increase alarmingly. Suppose we have eight-one equations and six unknowns, a modest problem by most standards. We would need to calculate over $3.4 \cdot 10^8$ solutions, the solution of each 6×6 subsystem requiring, if we use Gaussian elimination; $6^3/3$ arithmetic operations, for a total of $2.4 \cdot 10^{10}$ operations. Even a fast computer performing one arithmetic operation every microsecond would need nearly seven hours to find the L_1 solution. Problems with thousands of equations and dozens of unknowns would be impossible if we relied on the simple method. The fact that the author actually found the L_1 solution of an 81×6 system of double-precision numbers on a home computer without floating-point hardware demonstrates that a more efficient method exists.

Even if our system were sufficiently small that the simple method is feasible, another difficulty presents itself. Although **A**, a matrix of size $m \times n$, may be of rank n, to obtain a solution of Eq. (4.1) in the L_1 norm *every* $n \times n$ submatrix of **A** must be of rank n, that is, nonsingular. This condition, called the Haar condition, can be extremely restrictive. **A** usually comes from a mathematical model and **d** from experimental data. Sometimes the coefficients of a row of **A** remain the same while the data change slightly. When determining minor planet positions from a photographic plate, for example, we sometimes calculate the position twice, using different reference stars for each position. Because of different errors in the reference star positions, the calculated positions of the minor planet change but the coefficients in the mathematical model, such as the positions in space of the Earth and the minor planet at the moment of observation, remain the same. The Haar condition is violated. An L_1 algorithm, aside from being efficient, must also take into account the Haar condition. Should the third row of **A** of Eq. (6.14) be changed from (1 3) to (1 1) the Haar condition is violated—row three is the same as row 1—but the Barrodale and Robert's algorithm of Section 6.4 calculates, without difficulty, an L_1 solution of $X^T = (0.75 \ 0.25)$ with the sum of the absolute values of the residuals equal to 2.5.

The simple method for obtaining an L_1 solution is impractical. We might feel that a development for an L_1 solution paralleling that for a least squares solution of Eqns. (4.4)–(4.8) may be the way to proceed. Attempting this, however, we see that an L_1 solution has certain characteristic features not found in a least squares solution. The first jarring note comes upon differentiating the absolute value of a function and setting it to zero, as with Eq. (4.7). Take the function to be minimized as

$$f = \sum_{i=1}^{m} \left| \sum_{j=1}^{n} a_{ij} X_j - d_i \right|. \tag{6.15}$$

The derivative of the absolute value of a function, in general, is

$$f = |f(X)| \quad \Rightarrow \quad \frac{\partial f}{\partial X} = \frac{\partial f(X)}{\partial X} \mathrm{Sgn}(f(X)), \tag{6.16}$$

where $\mathrm{Sgn}(f(X))$ is the sign of $f(X)$:

$$\mathrm{Sgn}(f(X)) = \begin{cases} +1 & \text{if} \quad f(X) > 0, \\ 0 & \text{if} \quad f(X) = 0, \\ -1 & \text{if} \quad f(X) < 0. \end{cases} \tag{6.17}$$

If, for example, $f = |X|$, then

$$\frac{\partial f}{\partial X} = +1 \ \mathrm{Sgn}(X). \tag{6.18}$$

For positive X the derivative is a constant $+1$ and for negative X a constant -1.

But suppose X in Eq. (6.18) equals zero. The function $f = |X|$ has a point of discontinuity. The derivative does not exist at $X = 0$, although a right-directional derivative of $+1$ and a left-directional derivative of -1 exist. We assert, nevertheless, that Eq. (6.18) is zero at $X = 0$ because the sign function, being null at this point, multiplies out the singularity—the derivative not existing—and sets the equation to zero. The derivative of the absolute value of a function, $|f(X)|$, is zero, and hence a maximum or minimum, when $\mathrm{Sgn}(f(X)) = 0$.

Returning to Eq. (6.15), we differentiate it and set it to zero:

$$\frac{\partial f}{\partial X_j} = \sum_{i=1}^{m} a_{ij} \, \mathrm{Sgn} \left| \sum_{j=1}^{n} a_{ij} X_j - d_i \right|. \tag{6.19}$$

The difficulty now becomes clear; in general, Eq. (6.19) will not be zero. The sign function is a residual r_i. For Eq. (6.19) to be null, as with Eq. (4.7) of the least squares criterion, *all* of the residuals should be zero. But, because Eq. (4.1) is inconsistent, this cannot be true in general. Unlike the situation with least squares, the solution that minimizes the sum of the absolute values of the residuals does not lead to a vanishing gradient of Eq. (6.15). The L_1 solution of $\mathbf{X}^T = (0.5 \ 0.5)$ of Eq. (6.14), for example, has a gradient of $\partial f/\partial X_1 = 3$ and $\partial f/\partial X_2 = 11$. Nothing can be accomplished by setting, or attempting to set, the gradient (6.19) to zero.

This situation exemplifies a fundamental difference between the least squares and the L_1 criterion. The gradient of the overdetermined system with the least squares criterion is zero because it corresponds to an $n \times n$ system distinct from the original system: the normal equations do not coincide with any of the equations of the $m \times n$ overdetermined system nor does the solution of the normal equations satisfy any of the equations of the matrix $\mathbf{A}$. The L_1 criterion, on the other hand, satisfies n of the m equations, but it leads to a gradient that, consequently, cannot vanish.

Given that the simple method for obtaining an L_1 solution is inefficient and attempting to set the gradient to zero leads nowhere, what can be done? Numerous researchers, such as Armstrong and Godfrey (1979), have pointed out that the L_1 problem can be recast as a linear programming problem. Other approaches are possible—Abdelmalek (1971) attacks the L_1 problem by a series of iterations of a nonlinear algorithm based on the Davidon–Fletcher–Powell method described in the next chapter; and Bartels, Conn, and Sinclair (1978) develop what they call the direct descent method—but the Barrodale and Roberts algorithm (1974), widely used and available in many software packages, is based on linear programming and seems as efficient as any algorithm for the L_1 problem. The next section is devoted to a brief description of linear programming, and Section 6.4 to the Barrodale and Roberts algorithm.

6.3. Linear Programming

That the L_1 problem can be reduced to a linear programming problem is both fortunate and, in some respects, unfortunate. Fortunate, because the standard linear programming algorithm, Dantzig's simplex algorithm, is, aside from rare instances, far more efficient than the straightforward procedure for obtaining an L_1 solution mentioned in the previous section. With the simplex algorithm, solutions of problems with thousands of equations and hundreds of unknowns are possible. Unfortunately, many scientists, having little need of linear programming, are unfamiliar with its basic precepts. These, moreover, are expressed in terminology that sometimes confuses more than enlightens. The author agrees completely when Press, Flannery, Teukolsky, and Vetterling (1986, p. 313) write: "As you see, the subject of linear programming is surrounded by notational and terminological thickets. Both of these thorny defenses are lovingly cultivated by a coterie of stern acolytes who have devoted themselves to the field. Actually, the basic ideas of linear programming are quite simple."

The outline of linear programming in Press, Flannery, Teukolsky, and Vetterling's book is so clear that the author is almost inclined to refer the reader to their book and let it go at that. He, however, feels that the present book should be as self-contained as possible in case the reader has no easy access to the work referred to. (When once reading a biography of Tchaikovsky he was disconcerted to find that no analysis of that great composer's fifth symphony was given because the reader "necessarily" has a program with full analytical notes. Needless to say, the author did not have such a program.) And, besides, another version of a clear explanation, perhaps, better said, an attempted clear explanation, of linear programming should not be superfluous given the unfamiliarity with linear programming among many. A reader already acquainted with the subject may skip to Section 6.4; a reader desiring more details should refer to a standard work such as Gass (1975).

The basic problem of linear programming may be stated as:

maximize the linear objective function in n variables

$$f(X_1, X_2, \ldots, X_n) = C_1 X_1 + C_2 X_2 + \cdots + C_n X_n \tag{6.20}$$

subject to the primary constraints

$$X_1, X_2, \ldots, X_n \geq 0 \tag{6.21}$$

and the additional constraints

$$
\begin{aligned}
A_{11} X_1 + A_{12} X_2 + \cdots + A_{1n} X_n &\leq b_1, \\
A_{21} X_1 + A_{22} X_2 + \cdots + A_{2n} X_n &\leq b_n, \\
&\vdots \\
A_{m1} X_1 + A_{m2} X_2 + \cdots + A_{mn} X_n &\leq b_m,
\end{aligned}
\tag{6.22}
$$

with $b_1, b_2, \ldots, b_n \geq 0$. In matrix notation these equations are:

maximize

$$f(\mathbf{X}) = \mathbf{C}^T \cdot \mathbf{X} \tag{6.20'}$$

subject to

$$\mathbf{X} \geq 0 \tag{6.21'}$$

and

$$\mathbf{A} \cdot \mathbf{X} \leq \mathbf{b} \qquad (6.22')$$

with $\mathbf{b} \geq 0$. To minimize the objective function (6.21) we merely need to maximize $-\mathbf{C}^T \cdot \mathbf{X}$. The number of inequalities in Eq. (6.22), m, may be smaller than, equal to, or greater than n.

From Eq. (6.21) we note that the variables must be nonnegative. Why? Because the simplex method was originally devised for economic considerations with resource allocation, and the resources are positive or, at worst, zero. How many negative cars can a factory produce? The inequalities in Eq. (6.22) arise from the same considerations: the factory can spend up to some maximum amount, can use up so many raw products, can expend up to so many man hours, and so forth. Mathematically, this means an inequality, not an equality, although sometimes the inequalities may become equalities (hence the $\leq$ sign).

Geometrically, the constraints (6.21) and (6.22) define a convex polyhedron, known as a polytope, with a maximum of $n + m(n - 1)$ faces. The objective function (6.20) is a hyperplane in the n-dimensional space, its distance from the origin varying. Our mission is to maximize the distance of the hyperplane from the origin while at the same time satisfying the constraints. Conceptually, we start out in a positive orthant. Each constraint of the matrix $\mathbf{A}$ delimits a region beyond which the solution may not pass. After all of the constraints are imposed we are left with a region, the edges and interior of the polytope, where the solution must lie, unless the constraints are themselves inconsistent and no solution exists.

An example will make this more clear. Suppose that our problem is:

maximize

$$f(X_1, X_2) = X_1 + X_2 \qquad (6.23)$$

subject to

$$X_1 - X_2 \leq 3,$$
$$-X_1 + 2X_2 \leq 1,$$
$$X_1 \qquad \leq 4, \qquad (6.24)$$
$$X_2 \leq 2,$$

with X_1 and X_2 positive. Figure 6.2 shows the situation. The polytope is the figure O–A–B–C–D–E, and the solution must lie within or on the polytope, which has $2 + 4(2 - 1) = 6$ edges. The dashed lines show the objective function $X_1 + X_2$ that is to be maximized. The region within and on the edges of the polytope is referred to as the feasible region, and any solution in the region is a feasible solution.

The constraints may be such that no solution is feasible. If, for example, we were to specify two inequalities, and no others,

$$-X_1 - X_2 \leq -3,$$
$$X_1 + X_2 \leq 1, \qquad (6.25)$$

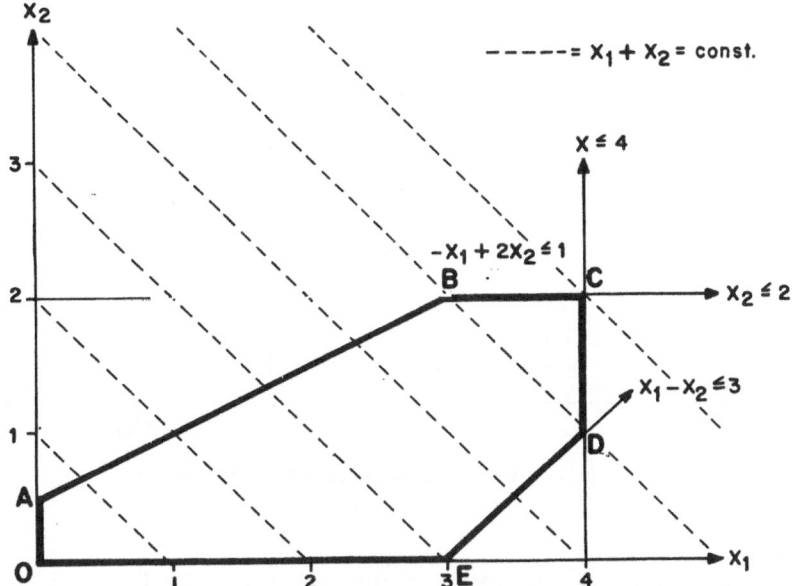

Figure 6.2. Polytope for Eq. (6.24), 0–A–B–C–D–E.

no feasible region exists. The first inequality translates to $X_1 + X_2 \geq 3$. The sum of X_1 and X_2 cannot be simultaneously greater than or equal to three and less than or equal to one. Or the solution may be unbounded. Strike the first three inequalities off Eq. (6.24). The feasible region extends to infinity along the X_1-axis. Should we eliminate the first two and the last inequalities, the feasible region would go to infinity along the X_2-axis.

We also appreciate that m has no special significance compared with n. A single inequality such as $X_1 + X_2 \leq 1$ defines, along with the primary constraints of course, a feasible region; in this instance, $m < n$. Strike the first two inequalities off Eq. (6.24) and we have a feasible region with $m = n$, and Eq. (6.24) defines a feasible region with $m > n$.

That the polytope is convex is one of the theorems of linear programming. Most scientists have an intuitive notion of convexity. A strict mathematical definition may be found in, among other references, Patterson (1969, p. 91). If E_n is an n-dimensional Euclidean space, a space whose definition of distance satisfies the properties of a vector norm given by Eqns. (2.12)–(2.14), and P_1 and P_2 are points in the space, then a convex linear subspace of E_n is a subset of E_n such that the point $\alpha_1 P_1 + \alpha_2 P_2$, with $\alpha_1, \alpha_2 \geq 0$ and $\alpha_1 + \alpha_2 = 1$, is also in E_n. If a convex linear subspace contains the points P_1 and P_2, it also contains all the points on the line uniting P_1 and P_2. Thus, in Figure 6.3, (a) is a convex linear subspace and (b) is not of the space E_3.

Another theorem of linear programming, which Gass (1975, Sec. 3.2) proves as he does the theorem of the previous paragraph, states that the optimal

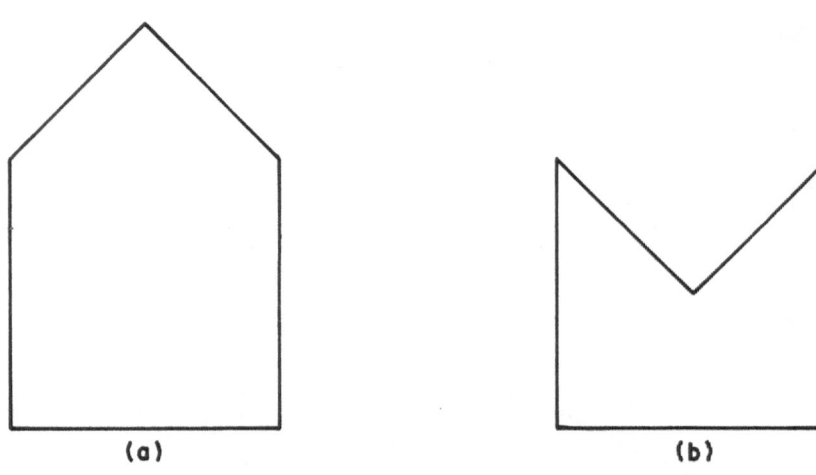

Figure 6.3. (a) Convex and (b) nonconvex linear subspaces.

feasible solution vector, the vector **X** that maximizes the linear objective function Eq. (6.20), corresponds to a point not in the interior of the polytope, but on one of the edges and generally at a vertex. In Figure 6.2, for example, the optimal solution of $\mathbf{X}^T = (4\ 2)$ is located at the vertex C and gives the maximum $f(\mathbf{X}) = 6$. At this point the distance between the origin and the line defined by the objective function is a maximum.

The theorem that the optimal feasible vector is found, generally, at a vertex, referred to because of its importance as the fundamental theorem of linear programming, may be proved in a number of ways. One way of proceeding, although less mathematically rigorous than Gass's proof, incorporates the pleasing characteristic of the intuitive properties of the gradient. An extremum of a function, usually a maximum or a minimum but occasionally a saddle point, occurs when the gradient vanishes. With a nonzero gradient we move in the direction indicated by the gradient to increase the function value, and in the opposite direction to minimize the function value. The gradient of Eq. (6.20) is a constant $\mathbf{C}^T$. If we start out, therefore, in the interior of the polytope, we can increase $f(X)$ by moving in the direction $\mathbf{C}^T$ until we hit an edge. The edge cannot be crossed because it delineates the boundary of the feasible region: crossing the edge the solution becomes infeasible. But the projection of the gradient, with the equation defined by the edge, indicates the new direction in which to proceed to maximize $f(\mathbf{X})$. Eventually, we end up at a vertex where $f(\mathbf{X})$, within the constraints defined by the inequalities, reaches its maximum. Occasionally, there may be more than one vertex with the same maximum, a degenerate solution.

As a concrete example, take Eqns. (6.23) and (6.24) and start at an interior point of the polytope, say $\mathbf{X}^T = (0.5\ 0.5)$, where $f(\mathbf{X}) = 1$. The gradient is a constant $\nabla f(\mathbf{X}) = (1\ 1)$, making an angle of $45°$ with respect to the X_1-,

X_2-axes and perpendicular to the contours defined by the objective function (6.23). Taking stepsizes of $\Delta\mathbf{X}^{\mathrm{T}} = (0.05 \ 0.05)$ and marching in the direction of the gradient, we finally bump against the edge $-X_1 + 2X_2 \leq 1$ at $\mathbf{X}^{\mathrm{T}} = (1.5 \ 1.25)$, where $f(\mathbf{X}) = 2.75$. At this point we move in the direction of the projected gradient, now making an angle of $26°.57$ with the X_1-axis instead of $45°$, until we reach the vertex B where $\mathbf{X}^{\mathrm{T}} = (3 \ 2)$ and $f(\mathbf{X}) = 5$. The edge $X_2 \leq 2$ cannot be crossed, and the projected gradient now has no X_2 component: X_2 cannot be increased more, but X_1 can. We move towards the vertex C, increasing $f(\mathbf{X})$. At C we achieve the maximum because a right-angle turn down the $X_1 \leq 4$ edge decreases the objective function.

In this example the solution is unique because the objective function (6.23) intersects the feasible region at only the point C where the distance from the origin is a maximum. But it may happen that the objective function coincides with an edge, giving rise to a degenerate solution. If the objective function is (6.21) but the only constraint is $X_1 + X_2 \leq 1$, the objective function coincides with the constraint and all pairs of points on the line $X_1 + X_2 = 1$ result in the same $f(\mathbf{X}) = 1$, although the points farthest from the origin are $\mathbf{X}^{\mathrm{T}} = (1 \ 0)$ and $\mathbf{X}^{\mathrm{T}} = (0 \ 1)$. The solution is degenerate.

If we remember that the L_1 method is closely related to linear programming, the discussion of the previous four paragraphs demonstrates what was contended in Section 6.2, that an L_1 solution may be degenerate with more than one solution vector giving the same sum of the absolute values of the residuals, and in Section 6.1, that n of the m equations are satisfied exactly. The latter follows from a linear programming solution being found at a vertex. At a vertex, n of the inequalities, including the primary constraints, are satisfied as equalities, which furnishes a thoroughly impractical way of solving a linear programming problem, for the same reasons as mentioned in Section 6.2. But to reiterate, solving every $n \times n$ subsystem would be, for reasonably sized m and n, horrendously time-consuming.

The first step of a practical linear programming algorithm, the simplex method developed by George Dantzig and associates in the late 1940s, consists of reducing the inequalities of Eq. (6.22) to equalities by the introduction of additional variables, called slack variables; there are m of these:

$$A_{11}X_1 + A_{12}X_2 + \cdots + A_{1n}X_n + X_{n+1} = b_1,$$
$$A_{21}X_1 + A_{22}X_2 + \cdots + A_{2n}X_n + X_{n+2} = b_2,$$
$$\vdots$$
$$A_{m1}X_1 + A_{m2}X_2 + \cdots + A_{mn}X_n + X_{n+m} = b_m; \tag{6.26}$$

in matrix notation

$$(\mathbf{A} \quad \mathbf{I})\begin{pmatrix} \mathbf{X} \\ \mathbf{X}_m \end{pmatrix} = \mathbf{b}, \tag{6.26'}$$

where $\mathbf{I}$ is an $m \times m$ identity matrix and $\mathbf{X}_m$ is the vector of slack variables. Let the partitioned matrix $(\mathbf{A} \ \mathbf{I})$ be denoted by $\mathbf{A}'$. The columns of $\mathbf{A}'$ are $n + m$

vectors in a space of m dimensions, m dimensions because the rank of $\mathbf{A}'$ is m. Of the $n + m$ vectors m are linearly independent and form a basis in the m space, and the remaining n vectors are nonbasic, expressable as linear combinations of the basic vectors. In linear programming terminology a basic vector is obtained by setting n of the $n + m$ variables to zero and solving for the remaining m. A solution given by a vector that is both basic and feasible is referred to, obviously enough, as a basic feasible solution.

The slack variables of Eq. (6.26) permit us to immediately have a basic feasible solution. Set $\mathbf{X} = 0$ and $\mathbf{X}_m = \mathbf{b}$. The nonbasic vector $\mathbf{X}$ coincides with the origin. The simplex method proceeds by a series of interchanges of basic and nonbasic vectors in such a way that the objective function is increased with each interchange. Our first concern, therefore, is: How do we exchange a basic and a nonbasic vector? Rather than present a general discussion, found in Section 3.3 of Gass's book (1975), we will give a concrete example, using Eqns. (6.23) and (6.24).

After slack varables are introduced these equations become:

maximize
$$X_1 + X_2 + 0X_3 + 0X_4 + 0X_5 + 0X_6 \tag{6.27}$$

subject to
$$
\begin{aligned}
X_1 &- X_2 + X_3 &&&&= 3, \\
-X_1 &+ 2X_2 &&+ X_4 &&= 1, \\
X_1 &&&&+ X_5 &&= 4, \\
&X_2 &&&&+ X_6 &= 2,
\end{aligned}
\tag{6.28}
$$

or, in matrix notation,

maximize
$$\mathbf{C}^{\mathrm{T}} \cdot \mathbf{X}, \tag{6.27'}$$

with
$$\mathbf{C}^{\mathrm{T}} = (1 \quad 1 \quad 0 \quad 0 \quad 0 \quad 0),$$

subject to
$$
\begin{pmatrix}
1 & -1 & 1 & 0 & 0 & 0 \\
-1 & 2 & 0 & 1 & 0 & 0 \\
1 & 0 & 0 & 0 & 1 & 0 \\
0 & 1 & 0 & 0 & 0 & 1
\end{pmatrix}
\begin{pmatrix}
X_1 \\ X_2 \\ X_3 \\ X_4 \\ X_5 \\ X_6
\end{pmatrix}
=
\begin{pmatrix}
3 \\ 1 \\ 4 \\ 2
\end{pmatrix}.
\tag{6.28'}
$$

The variables $X_3 \rightarrow X_6$ are the slack variables.

The rank of the coefficient matrix of Eq. (6.28) is four, as may be found in a number of ways. Perhaps the easiest to use is the singular value decomposition, discussed in Chapter 8, and find out how many nonzero singular values there are. If this is done we calculate the ordered singular values, to five decimals, as: 2.97558, 1.46489, 1.00000, 1.00000, 0, 0. Four nonzero singular values; rank four for the matrix. Four of the columns of the matrix, therefore, are independent and may be taken as the basic vectors, and the nonbasic vectors are expressed in terms of them.

Standard linear programming notation uses P_j for the columns of A' and P_0 for b. The vectors P_3 to P_6 of Eq. (6.28) are unitary vectors. (Notice that in Chapter 2 we used the symbol e_i for unitary vectors.) In Eq. (6.28) the initial basic feasible solution is simply $X^T = (0\ 0)$ and $X_m^T = (3\ 1\ 4\ 2)$. The solution, $X_1 = 0, X_2 = 0$, does not maximize the objective function, which as we already know has a maximum of six. Our initial basic feasible solution is not optimal. To proceed towards optimality we must move the solution from the origin to point C of Figure 6.2, either in one step, if possible, or by passing through a series of vertices such as O–E–D–C.

The coefficient matrix A' of Eq. (6.28) is of rank four and the columns P_3, P_4, P_5, and P_6 of A' constitute a unit matrix of rank four also. These vectors, therefore, are the basic vectors, and P_1 and P_2, the nonbasic vectors, may be expressed in terms of them by

$$P_1 = P_3 - P_4 + P_5;$$
$$P_2 = -P_3 + 2P_4 + P_6.$$

(6.29)

P_0 is also expressible by vectors in the basis:

$$P_0 = 3P_3 + P_4 + 4P_5 + 2P_6.$$

(6.30)

But we may interchange a basic and a nonbasic vector, reversing their roles. Suppose that we wish to introduce P_1 into the basis and remove one of the vectors P_3 to P_6. Subtract a multiple of P_1, call it θP_1, from P_0,

$$P_0 - \theta P_1 = 3P_3 + P_4 + 4P_5 + 2P_6 - \theta(P_3 - P_4 + P_5),$$

or

$$P_0 = \theta P_1 + (3 - \theta)P_3 + (1 + \theta)P_4 + (4 - \theta)P_5 + 2P_6$$

(6.31)

and inquire what conditions θ must fulfill to eliminate a vector in the basis. To have anything nontrivial happen θ must be nonzero. It must also be positive, otherwise, θP_1 will be negative, corresponding to a component of b being negative and contradicting the sense of the inequalities in Eq. (6.22). (A negative b_i may be made positive by multiplying both sides of the inequality by minus one, but this changes the less than or equal to sign to greater than or equal to.) Of the terms involving θ in Eq. (6.31) that are within parentheses, the term $1 + \theta$ lacks interest because it permits no basic vector to be removed. The term $4 - \theta$ allows the elimination of P_5, but drives P_3 negative. This leaves the term $3 - \theta$. If we set $\theta = 3$ the vector P_3 is removed from the basis and P_1 is introduced, or

$$P_0 = 3P_1 + 4P_4 + P_5 + 2P_6.$$

(6.32)

The new basis vectors, P_1, P_4, P_5, P_6, are mutually orthogonal like the old ones, with P_1 fulfilling the role of P_3, or $P_1^T = (1\ 0\ 0\ 0)$. From Eq. (6.29)

$$P_3 = P_1 + P_4 - P_5,$$
$$P_2 = -P_1 + P_4 + P_5 + P_6.$$

(6.33)

The remaining vectors are unchanged. Equation (6.28) becomes

$$
\begin{pmatrix}
1 & -1 & 1 & 0 & 0 & 0 \\
0 & 1 & 1 & 1 & 0 & 0 \\
0 & 1 & -1 & 0 & 1 & 0 \\
0 & 1 & 0 & 0 & 0 & 1
\end{pmatrix}
\begin{pmatrix}
X_1 \\ X_2 \\ X_3 \\ X_4 \\ X_5 \\ X_6
\end{pmatrix}
=
\begin{pmatrix}
3 \\ 4 \\ 1 \\ 2
\end{pmatrix},
\qquad (6.34)
$$

which gives the solution $\mathbf{X}^T = (3\ 0)$, $\mathbf{X}_m^T = (0\ 4\ 1\ 2)$, corresponding to point E of Figure 6.2.

Had we introduced $\mathbf{P}_2$ in the basis by substracting $\theta \mathbf{P}_2$ from $\mathbf{P}_0$ we would have ended up at point A of Figure 6.2. We could repeat the procedure and move from point E to point D or from A to B. In adding and removing vectors to the basis we must keep in mind the restrictions on θ. If a suitable θ cannot be found, the interchange cannot be made. To continue the example, if we wish to introduce $\mathbf{P}_2$ into the basis we find, from Eqns. (6.32) and (6.33), that

$$
\mathbf{P}_0 = \theta \mathbf{P}_2 + (3 + \theta)\mathbf{P}_1 + (4 - \theta)\mathbf{P}_4 + (1 - \theta)\mathbf{P}_5 + (2 - \theta)\mathbf{P}_6. \quad (6.35)
$$

The only possibility for θ is $\theta = 1$, which interchanges $\mathbf{P}_2$ and $\mathbf{P}_5$ in the basis and gives the system

$$
\begin{pmatrix}
1 & 0 & 0 & 0 & 1 & 0 \\
0 & 0 & 2 & 1 & -1 & 0 \\
0 & 1 & -1 & 0 & 1 & 0 \\
0 & 0 & 1 & 0 & -1 & 1
\end{pmatrix}
\begin{pmatrix}
X_1 \\ X_2 \\ X_3 \\ X_4 \\ X_5 \\ X_6
\end{pmatrix}
=
\begin{pmatrix}
4 \\ 3 \\ 1 \\ 1
\end{pmatrix},
\qquad (6.35)
$$

with solution $\mathbf{X}^T = (4\ 1)$, $\mathbf{X}_m^T = (0\ 3\ 0\ 1)$; this corresponds to point D of Figure 6.2.

The transformations involved in adding and removing vectors from the basis are actually the same as those for the solution of a linear system by Gaussian elimination given in Section 4.4. In going from Eq. (6.28) to Eq. (6.34) the column $\mathbf{P}_1$ is reduced to a unitary vector and the remaining columns transformed by premultiplication by the matrix

$$
\mathbf{M}_1 =
\begin{pmatrix}
1 & 0 & 0 & 0 \\
1 & 1 & 0 & 0 \\
-1 & 0 & 1 & 0 \\
0 & 0 & 0 & 1
\end{pmatrix}. \qquad (6.36)
$$

The main difference between forming the matrix $\mathbf{M}_i$ here and in Section 4.4 is that here we are uninterested in reducing a given matrix to upper triangular form; the rule for selecting the pivot element is, therefore, different. In adding and removing vectors from the basis the pivot is selected so that the transformation matrix $\mathbf{M}_i$ corresponds to a θ that satisfies the restrictions mentioned previously.

To understand how the pivot is selected we now reintroduce a bit of generality. Equation (6.26) may be written in terms of the so-called simplex table as

P_0	P_1	P_2	$\cdots$	P_j	$\cdots$	P_n	P_{n+1}	P_{n+2}	$\cdots$	P_{n+m}
b_1	a_{11}	a_{12}		a_{1j}		a_{1n}	1	0		0
b_2	a_{21}	a_{22}		a_{2j}		a_{2n}	0	1	$\cdots$	0
$\vdots$	$\vdots$	$\vdots$		$\vdots$		$\vdots$	$\vdots$	$\vdots$		$\vdots$
b_m	a_{m1}	a_{m2}		a_{mj}		a_{mn}	0	0		1

$$(6.37)$$

In a given column P_j there are m possible pivots

$$\theta_i = \frac{b_i}{a_{ij}} \quad (a_{ij} \neq 0) \tag{6.38}$$

and to satisfy the restrictions on θ we select as pivot the element a_{kj} that corresponds to $\theta_k = \min \theta_i$.

Rather than explicitly form the matrix $\mathbf{M}_i$, we more efficiently transform the simplex table by working with the column P_0, the n nonbasic columns that will be modified, and the one column that will be removed from the basis and, if a_{kl} is the pivot element:

(1) Divide the pivot row k by the pivot element,

$$a_{kj} = \frac{a_{kj}}{a_{kl}}. \tag{6.39}$$

(2) For the pivot column l set

$$a'_{il} = \delta_{ik}, \tag{6.40}$$

where δ_{ik} is the Kronecker delta.

(3) Subject the remaining elements to the transformation

$$a'_{ij} = a_{ij} - a_{il} a'_{kj}. \tag{6.41}$$

(For the column P_0 substitute b_i for a_{ij}.)

Here is one more example that matrix notation, such as Eq. (6.36), leads to inefficient computations if applied directly. Actually, efficient algorithms for the simplex method do not assign storage for the basic vectors at all, but keep track of them by suitable indexing and render unnecessary the second step of the previous algorithm. With these steps we need $n + 1$ divisions plus $m(n + 1)$ multiplications for an operation count of the order of mn, whereas the operation count would be mn^2 if the matrix $\mathbf{M}_i$ were multiplied into the simplex table.

If the nonbasic columns of $\mathbf{A}'$ are also sparse, as sometimes happens, then we can take advantage of special sparse matrix techniques for even greater economy of storage and execution efficiency. See Tewarson (1973, Chap. 8) for details.

The discussion so far has shown how vectors may be added and removed from the basis. But if this were done in an unsystematic way, the simplex algorithm would hardly be efficient. On the contrary, it would be inefficient and consist of moving haphazardly from one vertex to another and checking for a maximum of the objective function. The simplex algorithm proceeds in a more systematic manner by introducing some auxiliary variables

$$Z_j = \sum_{i=1}^{m} a_{ij} C_i, \qquad j = 1, \ldots, n, \tag{6.42}$$

where C_i is the C_i of the objective function associated with a basis vector $\mathbf{P}_i$. (This may sound a little cryptic; it will be more clear in a moment.) Gass (1975, Sec. 4.1) demonstrates that if for any basic feasible solution the conditions $Z_j - C_j \geq 0$ are fulfilled for all $j = 1, \ldots, n$, then the objective function has reached its maximum value. (Gass actually refers to the conditions $Z_j - C_j \leq 0$ because he minimizes the objective function. In linear programming we can always switch from a maximization to a minimization problem by changing the sign of the objective function.) The simplex algorithm examines each $Z_j - C_j$ and takes, as the column to be introduced into the basis, that which has the largest $Z_j - C_j$.

To see how this works in practice we go back to Eq. (6.28) written in the form of a simplex table:

	C	$\mathbf{P}_0$	$\mathbf{P}_1$	$\mathbf{P}_2$	$\mathbf{P}_3$	$\mathbf{P}_4$	$\mathbf{P}_5$	$\mathbf{P}_6$	θ
$\mathbf{P}_3$	0	3	①	-1	1	0	0	0	$3/1 = 3$
$\mathbf{P}_4$	0	1	-1	2	0	1	0	0	—
$\mathbf{P}_5$	0	4	1	0	0	0	1	0	$4/1 = 4$
$\mathbf{P}_6$	0	2	0	1	0	0	0	1	
		0	-1	-1	0	0	0	0	

The first column shows the vectors currently in the basis and their associated C's. The Z's of Eq. (6.42) are given in the last row and are formed by multiplying each C in the second column by the elements in column $\mathbf{P}_j$. For example, for column $\mathbf{P}_1$ we have, given that $C_1 = 1$, $C_2 = 1$, $C_3 = C_4 = C_5 = C_6 = 0$, $Z_1 = 0 \cdot 1 + 0 \cdot (-1) + 0 \cdot 1 + 0 \cdot 0 - C_1 = -1$. The element in the last row of the column $\mathbf{P}_0$ is the current value of the objective function. The column labeled θ shows the possible values for the pivot, and the circled element in the pivot column is the current pivot. Actually, because there are two negative Z's and both are equal we could just a well start out by bringing $\mathbf{P}_2$, rather than $\mathbf{P}_1$, into the basis. $\mathbf{P}_1$ is selected here to replace $\mathbf{P}_3$ in the basis.

After the necessary transformations we have the following simplex table:

	C	P_0	P_1	P_2	P_3	P_4	P_5	P_6	θ
P_1	1	3	1	−1	1	0	0	0	—
P_4	0	4	0	1	1	1	0	0	$4/1 = 4$
P_5	0	1	0	①	−1	0	1	0	$1/1 = 1$
P_6	0	2	0	1	0	0	0	1	$2/1 = 2$
		3	0	−2	1	0	0	0	

The value of $Z_2 = -2$ indicates that P_2 should be brought into the basis to replace P_3.
This results in:

	C	P_0	P_1	P_2	P_3	P_4	P_5	P_6	θ
P_1	1	4	1	0	0	0	1	0	—
P_4	0	3	0	0	2	1	−1	0	$3/2 = 1.5$
P_2	1	1	0	1	−1	0	1	0	—
P_6	0	1	0	0	①	0	−1	1	$1/1 = 1$
		5	0	0	−1	0	2	0	

Both P_1 and P_2 are in the basis, but the negative Z for P_3 indicates that it should be introduced as a basis vector and P_6 removed.
We end up with:

	C	P_0	P_1	P_2	P_3	P_4	P_5	P_6	θ
P_1	1	4	1	0	0	0	1	0	—
P_4	0	1	0	0	0	1	1	−2	—
P_2	1	2	0	1	0	0	0	1	—
P_3	0	1	0	0	1	0	−1	1	—
		6	0	0	0	0	1	1	

All Z's are positive, indicating that the maximum has been reached. The maximum of six is found under the P_0 column. Both P_1 and P_2 are in the basis, and we calculate $\mathbf{X}^T = (4\ 2)$, $\mathbf{X}_m^T = (1\ 1\ 0\ 0)$. Therefore, $X_1 = 4, X_2 = 2$ maximizes the objective function.

The simplex algorithm gains efficiency by taking as the pivot column the one with most negative Z. But how efficient is it? Under certain circumstances the algorithm may behave badly and examine nearly all of the vertices. In other words, the algorithm may conceivably require a factorial number of arithmetic operations. But this seldom occurs in practice. The algorithm usually terminates with between m and $n + m$ passes, where pass is defined as one reduction of the simplex table. Three passes sufficed in the simple example above.

Degeneracy may occur and when it does the objective function does not increase with each pass. In theory, there is no guarantee that the simplex algorithm will converge to a solution giving a maximum of the objective function in the presence of degeneracy, but in practice degeneracy poses no problems.

The elements of linear programming given so far are sufficient for an understanding of the L_1 algorithm, but for the sake of completeness the author would like to comment on some additional features of linear programming.

We have assumed that the elements of the vector $\mathbf{b}$ of Eq. (6.22) are positive, or at least nonnegative. Such a restriction is unnecessary as a negative component of b can be made positive by multiplying both sides of the inequality by minus one. But this changes the sense of the inequality from less than or equal to to greater than or equal to, with the complication that the origin is now excluded as an initial basic feasible solution. The attentive reader may have noticed that Eq. (6.25), the example of inconsistent inequalities, does not conform to the assumptions of Eq. (6.22) because after multiplication by minus one the inequality reverses itself.

The technique for dealing with negative components of $\mathbf{b}$ is to introduce new variables, called artificial variables, and use them to find an initial basic feasible solution to the original problem. Then the simple algorithm proceeds as outlined above. See Gass (1975, Sec. 4.3) for details. With the L_1 algorithm, the introduction of artificial variables is unnecessary because, as we shall see, an initial basic feasible solution is immediately available.

Every linear programming problem like Eqns. (6.20)–(6.22) with n unknowns and m constraints, not including the primary constraints, has another formulation with m unknowns and n constraints:

minimize $\quad\quad\quad\quad\quad\quad \mathbf{b}^\mathrm{T} \cdot y \quad\quad\quad\quad\quad\quad\quad$ (6.43)

subject to

$$\mathbf{A}^\mathrm{T} \cdot y \geq \mathbf{C}^\mathrm{T} \quad\quad\quad\quad\quad\quad (6.44)$$

and

$$y \geq 0. \quad\quad\quad\quad\quad\quad (6.45)$$

Equations (6.20)–(6.22) are referred to as the primal problem and Eqns. (6.43)–(6.45) as the dual problem. If the primal problem has a solution, so does the dual, and the dual solution may sometimes be computationally less demanding to determine.

Although the simplex algorithm is the one most often used for linear programming problems, other approaches are possible. Karmarkar (1984) transforms the primal linear programming program to the canonical form

minimize $\quad\quad\quad\quad\quad\quad \mathbf{C}^\mathrm{T} \cdot \mathbf{X} \quad\quad\quad\quad\quad\quad\quad$ (6.46) subject to

$$\mathbf{A} \cdot \mathbf{X} = 0, \tag{6.47}$$

$$\mathbf{X} \geq 0, \tag{6.48}$$

$$\sum_{i=1}^{n} X_i = 1. \tag{6.49}$$

Equations (6.48) and (6.49) correspond to a simplex, a geometrical figure whose rigorous definition we give in the next chapter. For the moment we can say that in a space of n dimensions a simplex is a regular polyhedron with $n + 1$ vertices. In a two-dimensional space, for example, a simplex is a triangle.

Rather than jump from one vertex of a polytope to another, Karmarkar's algorithm starts from an initial basic feasible solution in the center of the simplex, and decreases the value of the objective function with each iteration by monitoring a potential function. Unlike the simplex algorithm which in theory, but hardly ever in practice, may exhibit factorial behavior, Karmarkar's algorithm never exceeds polynomial time execution. Karmarkar feels that his algorithm compares favorably with the simplex algorithm, and for parallel computing may be superior, but more computing experience is required. Karmarkars's article is rather difficult to follow—at least the author found it so—but Franklin (1987) presents a clear discussion of the algorithm and its convergence properties.

6.4. The L_1 Algorithm and Error Estimates

Numerous algorithms have been proposed for L_1 estimation, but that of Barrodale and Roberts (1973) appears to be the one most widely used. Barrodale and Roberts use a modification of the primal simplex algorithm to obtain the L_1 solution. A competing algorithm by Abdelmalek (1975) is just as efficient as the Barrodale and Roberts algorithm, but apparently less widely used. Abdelmalek employs a modification of the dual simplex algorithm. Armstrong and Godfrey (1979) demonstrate that Abdelmalek's algorithm is equivalent to Barrodale and Roberts. Bartels, Conn, and Sinclair (1978) developed the direct descent method, which does not rely on linear programming. The direct descent method, apparently no more efficient than the linear programming method, seems less frequently used than the popular Barrodale and Roberts algorithm, the only one we will discuss for L_1 estimation.

An L_1 solution may be obtained from linear programming. Let u_i be a positive residual and v_i a negative residual. Express a component X_j of the solution vector $\mathbf{X}$ as the difference of two nonnegative variables f_j and g_j, $X_j = f_j - g_j$. Then the problem of L_1 estimation corresponds to an optimal solution of the primal linear programming problem:

minimize $$\sum_{i=1}^{m} (u_i + v_i) \tag{6.50}$$

subject to

$$\sum_{j=1}^{n} (f_j - g_j)a_{ij} + u_i - v_i = d_i, \qquad i = 1, \ldots, m, \qquad (6.51)$$

and

$$f_j, g_j, u_i, v_i \geq 0. \qquad (6.52)$$

Equations (6.50)–(6.52) may be entered into a simplex table similar to the one used in the previous section:

	C	$\mathbf{d}$	$\mathbf{b}_1 \cdots \mathbf{b}_n$		$\mathbf{g}_1 \cdots \mathbf{g}_n$		$\mathbf{u}_1 \cdots \mathbf{u}_m$	$\mathbf{v}_1 \cdots \mathbf{v}_m$
$\mathbf{u}_1$	1	d_1	a_{11}	a_{1n}	$-a_{11}$	$-a_{1n}$	1 0	-1 0
$\mathbf{u}_2$	1	d_2	a_{21}	a_{2n}	$-a_{21}$	$-a_{2n}$	0 0	0 0
$\vdots$	1	$\vdots$	$\vdots$	$\vdots$	$\vdots$	$\vdots$	$\vdots$ $\vdots$	$\vdots$ $\vdots$
$\mathbf{u}_m$	1	d_m	a_{m1}	a_{mn}	$-a_{m1}$	$-a_{mn}$	0	0 -1
		$\sum_i d_i$	$\sum_i a_{i1}$	$\sum_i a_{in}$	$-\sum_i a_{i1}$	$-\sum_i a_{in}$	0 0	-2 -2

The bottom row show the starting values for the variables Z_j of Eq. (6.42), which Barrodale and Roberts call marginal costs. $\mathbf{d}$ corresponds with $\mathbf{P}_0$, $\mathbf{u}_j$ and $\mathbf{v}_j$ are columns of slack variables, and $\mathbf{f}_j$ and $\mathbf{g}_j$ are the nonbasic vectors. At first sight it appears as if the simplex table involves too many variables, but from Eq. (6.51), if a positive residual u_i is present, v_i must be zero and vice versa. Therefore, $u_i = -v_i$. In fact, because $\mathbf{u}_i = u_i \mathbf{e}_i$, where $\mathbf{e}_i$ is the unit vector of Section 2.3, we can write $\mathbf{u}_i = -\mathbf{v}_i$. By a similar line of reasoning we deduce that $\mathbf{b}_j = -\mathbf{g}_j$. (If $\mathbf{a}_{i,j}$ denotes the jth column of the matrix A' of Eq. (6.26), then $\mathbf{f}_j = f_j \mathbf{a}_{i,j}$.) Much of the information in the table is, therefore, redundant. Only the vectors $\mathbf{d}, \mathbf{f}, \ldots, \mathbf{f}_n$ need be stored explicitly. Furthermore, an initial basic feasible solution is immediately available: $f_1, \ldots, f_n = 0$, $u_1 = d_1, \ldots,$ $u_m = d_m$; if a d_i is negative, change the sign of that row and replace $\mathbf{u}_i$ in the basis by $\mathbf{v}_i$. We need not bother with artificial variables.

The primal simplex algorithm could be applied to the table, performing a transformation whenever a marginal cost Z_j is positive. Positive, rather than negative, because we are minimizing, not maximizing, the objective function. But the nonnegativity constraints on the vectors $\mathbf{f}_j, \mathbf{g}_j, \mathbf{u}_j, \mathbf{v}_j$, a somewhat artificial restriction, lead to a certain inefficiency. Barrodale and Roberts modify the simplex algorithm slightly to obtain greater efficiency.

The Barrodale and Roberts L_1 algorithm is implemented in two stages. Stage 1 refers to the first n iterations and stage 2 to subsequent iterations. In stage 1 the vector to enter the basis is taken as the vector $\mathbf{f}_j$ or $\mathbf{g}_j$ with the largest nonnegative marginal cost. Because the marginal cost of $\mathbf{g}_j$ is the negative of that for $\mathbf{f}_j$ it need not be calculated separately. In stage 2 the vector to enter the basis is chosen from among the nonbasic $\mathbf{u}_j$ and $\mathbf{v}_j$ as the one with the greatest positive marginal cost. If no such vector can be found, all marginal costs are nonpositive and the solution has been obtained.

In both stages the vector to leave the basis is the basic vector $\mathbf{u}_j$ or $\mathbf{v}_j$ that causes the maximum reduction in the objective function. This is the modification to the simplex algorithm that permits greater efficiency. In the standard simplex algorithm the pivot must be selected according to the rule that $\theta_k = \min \theta_i$ (see Eq. (6.38)) or a negative vector will be introduced into the basis (see Eq. (6.31)) which is not allowed. We cannot guarantee that the vector to leave the basis causes the maximum reduction in the objective function. The L_1 algorithm allows greater flexibility. Once a pivot, found in the normal way, is at hand, rather than proceed to the simplex transformation (Eqns. (6.39)–(6.41)), we subtract twice the value of the pivot from the marginal cost of the pivot column. If the result is nonpositive the simplex transformation is applied. Otherwise, twice the pivot row is subtracted from the marginal cost row, the pivot row is multiplied by minus one, and u_j or v_j in the basis corresponding to the pivot row is replaced by v_j or u_j. Then a new pivot is found and the procedure repeated until a simplex transformation must of necessity be performed. In this manner we pass through several polytope vertices before having to calculate a simplex transformation.

This trick is possible in the L_1 algorithm, and not the simplex algorithm, because both vectors u_j and v_j, the sum of whose marginal costs is always minus two, are available. If one of them is driven negative it can be replaced by the other, which will not be negative as long as twice the pivot, when subtracted from the marginal cost of the pivot column, does not exceed minus two. The net result is a substantial decrease in the number of simplex transformation required.

The final table may contain the negative basic vector f_j or g_j. Any such row should be multiplied by minus one and the negative vector b_j or g_j replaced by g_j or f_j.

An example illustrates the preceding discussion. Suppose we want the L_1 solution of the 3×2 system

$$\begin{pmatrix} 1 & -1 \\ 1 & 1 \\ 2 & 1 \end{pmatrix} \begin{pmatrix} X_1 \\ X_2 \end{pmatrix} = \begin{pmatrix} 2 \\ 4 \\ 8 \end{pmatrix}. \tag{6.53}$$

This simple system may be solved by the direct approach of considering all combinations of equations. Solving the first and the third equation gives the L_1 solution of $\mathbf{X}^T = (3.33333 \quad 1.33333)$ with $\mathbf{r}^T = (0 \quad 0.66667 \quad 0)$.

To apply the Barrodale and Roberts algorithm the data from Eq. (6.53) are entered into the simplex table. For the purpose of illustration we show the complete table, with no vectors suppressed.

	C	d	f_1	f_2	g_1	g_2	u_1	u_2	u_3	v_1	v_2	v_3
u_1	1	2	①	-1	-1	1	1	0	0	-1	0	0
u_2	1	4	1	1	-1	-1	0	1	0	0	-1	0
u_3	1	8	2	1	-2	-1	0	0	1	0	0	-1
		14	4	1	-4	-1	0	0	0	-2	-2	-2

Because of having the largest positive marginal cost, f_1 becomes the pivot column. The pivot element is circled. Twice the pivot element subtracted from the marginal cost gives the positive result of two. Therefore, instead of applying the standard simplex transformation, we interchange u_1 and v_1 in the basis by the technique already presented, giving:

	C	d	f_1	f_2	g_1	g_2	u_1	u_2	u_3	v_1	v_2	v_3
v_1	1	-2	-1	1	1	-1	-1	0	0	1	0	0
u_2	1	4	①	1	-1	-1	0	1	0	0	-1	0
u_3	1	8	2	1	-2	-1	0	0	1	0	0	-1
		10	2	3	-2	-3	-2	0	0	0	-2	-2

Notice that the value of the objective function has been reduced from fourteen to ten without our having performed a simplex transformation.

At this point f_2 has a larger marginal positive cost than f_1, but we may not switch our attention to it. We must stay with the same column until it enters the basis. There is a choice of two pivot elements. But neither of them allows the expedient of reducing the objective function by interchanging a u_i with a v_i in the basis. The circled element selected as the pivot when multiplied by two and subtracted from the marginal cost gives zero, which is nonpositive and we proceed with the standard simplex transformation and bring f_1 into the basis to replace u_2:

	C	d	f_1	f_2	g_1	g_2	u_1	u_2	u_3	v_1	v_2	v_3
v_1	1	2	0	②	0	-2	-1	1	0	1	-1	0
f_1	0	4	1	1	-1	-1	0	1	0	0	-1	0
u_3	1	0	0	-1	0	1	0	-2	1	0	2	-1
		2	0	1	0	-1	-2	-2	0	0	0	-2

The pivot column is now f_2. No interchanges of u_i's and v_i's are possible. The next simplex transformation results in:

	C	d	f_1	f_2	g_1	g_2	u_1	u_2	u_3	v_1	v_2	v_3
f_2	1	1	0	1	0	-1	$-\frac{1}{2}$	$\frac{1}{2}$	0	$\frac{1}{2}$	$-\frac{1}{2}$	0
f_1	1	3	1	0	-1	0	$\frac{1}{2}$	$\frac{1}{2}$	0	$-\frac{1}{2}$	$-\frac{1}{2}$	0
u_3	1	1	0	0	0	0	$-\frac{1}{2}$	$-\frac{3}{2}$	1	$\frac{1}{2}$	$\frac{3}{2}$	-1
		1	0	0	0	0	$-\frac{3}{2}$	$-\frac{5}{2}$	0	$-\frac{1}{2}$	$\frac{1}{2}$	-2

f_2 has replaced u_2 in the basis, but v_2 still has a positive marginal cost. Applying the simplex transformation to bring this vector into the basis, we obtain:

	C	$\mathbf{d}$	$\mathbf{f_1}$	$\mathbf{f_2}$	$\mathbf{g_1}$	$\mathbf{g_2}$	$\mathbf{u_1}$	$\mathbf{u_2}$	$\mathbf{u_3}$	$\mathbf{v_1}$	$\mathbf{v_2}$	$\mathbf{v_3}$
$\mathbf{f_2}$	0	$1\frac{1}{3}$	0	1	0	-1	$-\frac{2}{3}$	0	$\frac{2}{3}$	$\frac{2}{3}$	0	$-\frac{2}{3}$
$\mathbf{f_1}$	0	$3\frac{1}{3}$	1	0	-1	0	$\frac{1}{3}$	0	$\frac{1}{3}$	$-\frac{1}{3}$	0	$-\frac{1}{3}$
$\mathbf{v_2}$	0	$\frac{2}{3}$	0	0	0	0	$-\frac{1}{3}$	-1	$\frac{2}{3}$	$\frac{1}{3}$	1	$-\frac{2}{3}$
	1	$\frac{2}{3}$	0	0	0	0	$-\frac{4}{3}$	-2	$-\frac{1}{3}$	$-\frac{2}{3}$	0	$-\frac{5}{3}$

No marginal cost is positive; we have arrived at the solution. The value of the objective function coincides with what the direct approach gives,

$$\sum_{i=1}^{3} |r_i| = 0.66667,$$

as does the solution, $\mathbf{f_1} = 3\frac{1}{3}$ and $\mathbf{f_2} = 1\frac{1}{3}$.

This example has been somewhat atypical in that three simplex transformations were necessary. But a straight application of the simplex method to the original table also requires three transformations, as the reader may verify. This is an instance of the efficiency of an algorithm only becoming apparent when it is applied to a nontrivially sized problem, which a 3 × 2 system is not. The Barrodale and Roberts article (1973) mentions an example of a system of size 201 × 2. The standard simplex algorithm required 201 iterations to calculate an L_1 solution whereas their algorithm did it in seven iterations.

Barrodale and Roberts published a FORTRAN subroutine for their algorithm (1974). The subroutine, written in FORTRAN-IV, incorporates the abundance of GOTO statements characteristic of unstructured languages, like early versions of FORTRAN and BASIC. Out of 167 lines of code, not counting comments, GOTO occurs thirty-two times. In the section of the subroutine that determines the vector to leave the basis, 35% of the statements are GOTOs or contain GOTO. The resulting spaghetti code is rather difficult to follow, even with the aid of a flowchart.

Figure 6.4 presents a version of the Barrodale and Roberts algorithm, given as a complete program rather than a subroutine, that the author has translated to FORTRAN-77. This release of FORTRAN includes many control structures, such as the IF–THEN–ELSE and DO WHILE lacking in FORTRAN-IV, that permit eliminating most of the GOTOs. Less than 3% of the code in Figure 6.4 involves the GOTO. The remaining six GOTOs could only be removed by a complete restructuring of the program, but such an effort hardly seems worthwhile just for six GOTOs.

The Barrodale and Roberts algorithm calculates a solution even if the coefficient matrix is subrank and tests for possible solution degeneracy. The program in Figure 6.4 determines the rank of the coefficient matrix, but omits the degeneracy test because degeneracy is unlikely for problems taken from the real world. (Should we wish to include the test, we merely need to examine the value of the element $A(M + 2, N + 1)$. A value of zero means possible degeneracy, one normal termination, and two premature terminations caused by rounding errors. The latter condition should never occur for well-posed problems, of which we say more in Chapter 8.) The program calculates the solution, the residuals, the sum of the absolute values of the residuals, and the number of iterations performed. The user enters the coefficient matrix and the right-hand side interactively. For large problems we would more likely enter these data from a data file on disk or tape.

How does the L_1 method compare with the method of least squares in terms of efficiency? One great advantage of the latter is that the coefficient matrix A of size $m \times n$ may be compressed to the size $n \times n$ or even, because of the symmetry of the normal equations, to size $n(n + 1)/2$. Even orthogonal transformations may be applied sequentially in an array of size considerably smaller than $m \times n$. The L_1 algorithm of Figure 6.4, unfortunately, works with the full $m \times n$ matrix A, a disadvantage when $m \gg n$.

In terms of operation count the method of least squares, if implemented by the formation of normal equations and their Cholesky decomposition, requires, as Section 5.2 shows, $mn^2/2 + n^3/6$ operations. The L_1 method, as implemented by linear programming, needs mn operations for each reduction of the simplex table and between m and $n + m$ reductions. But the Barrodale and Roberts algorithm is more efficient than a direct application of the simplex algorithm. Based on the author's experience, the Barrodale and Roberts algorithm terminates after a number of iterations proportional to n, perhaps $10n$ to give a rough estimate. The operations count would then be $10mn^2$, higher than that for least squares. But one application of the L_1 algorithm furnishes a robust solution whereas least squares must be iterated before achieving robustness; between perhaps ten and twenty times for iteratively reweighted least squares. If we take into account this difference, the operation counts for the two methods are roughly comparable.

But the operation count is not the only consideration. If m is sufficiently large that the matrix A does not fit into the computer's available physical memory, the L_1 algorithm generates high paging rates and inefficient CPU use on virtual memory systems. The algorithm will not work at all on computers without virtual memory if physical memory is exceeded, unless we write an overlay program or use tape or disk for storing intermediate results, bound to be inefficient. The author's personal experience indicates that if A fits into the available physical memory the L_1 algorithm competes well with least squares. But if A exceeds physical memory by a significant factor—most virtual memory systems handle a slightly oversized A without serious performance degradation—the L_1 algorithm becomes inefficient because of high paging rates; iteratively reweighted least squares then produces a robust solution with less labor.

As a concrete example the author once solved a system of size $21{,}365 \times 41$ in double-precision on a VAX-11/780 with 1 MB of physical memory by both the L_1 algorithm and iteratively reweighted least squares. Ten iterations of a least squares algorithm were sufficient for a robust solution. The L_1 algorithm required 341 reductions of the simplex table. From just looking at the operation counts, $1.8 \cdot 10^8$ for least squares and $2.9 \cdot 10^8$ for the L_1 algorithm, we expect that the latter would need 60% more CPU time. But the L_1 algorithm actually used up three times more CPU time. The matrix **A** exceeded physical memory by a factor of seven, guaranteeing high paging rates.

On another occasion, however, the author solved a 2,064 × 30 system in double-precision on an IBM 4341. **A** comfortably fits into the partition of 600 KB of the 4 MB of the machine's physical memory. The L_1 solution actually required less time than iteratively reweighted least squares.

Which of the two algorithms is more efficient, L_1 or iteratively reweighted least squares, depends on the size of the problem and the computer available. We cannot make a definite recommendation in favor of one or the other of the two algorithms.

Before leaving the L_1 algorithm we should address the question of how to obtain error estimates for the components of the solution vector **X**. The method of least squares furnishes, in the variance–covariance matrix discussed in Section 5.3, a convenient and computationally not too expensive way to estimate the errors. Can we do something similar in the L_1 method, where the variance–covariance matrix does not arise?

Perhaps the simplest way, but computationally demanding, is to use Monte Carlo simulation (Press, Flannery, Teukolsky, and Vetterling, 1986, pp. 529–532). To the vector **d** of Eq. (4.1) add random noise with the distribution of Eq. (6.4). Because the absolute value sign makes it difficult to implement the usual inverse mapping of random digits between zero and one—a reader unfamiliar with random number generation may find Chapter 10 of Forsythe, Malcolm, and Moler (1977) useful—the random noise with the L_1 distribution may be calculated by the acceptance–rejection method (Zelen and Severo, 1965, p. 925). Then calculate an L_1 solution. Repeat the process with different noise added to **d**. Do this a number of times, say ten, and use the dispersions in the components of the solution as the error estimates. For work in the L_1 norm the dispersion should be the MAD of Eq. (4.12) generalized to a system of m equations and n unknowns,

$$\text{MAD} = \sum_{i=1}^{m} \frac{|r_i|}{m - n}. \tag{6.54}$$

Monte Carlo simulation is time-consuming (especially when $m \gg n$, with m and n large), and calculating a single solution, let alone ten, taxes one's computer.

The author developed a method for obtaining error estimates with L_1 solutions that is computationally more efficient than Monte Carlo simulation (Branham, 1986). Section 5.3 derived formulas for least squares error estimates, which the formulas given here for L_1 error parallel closely. Start with Eq. (4.1) and assume that the errors in **X** depend only on **d**. Let $\varepsilon_{i,j}$ be the error in ith component of **X** caused by the jth component of **d**. Then from elementary calculus

$$\varepsilon_{i,j} = \frac{\partial X_i}{\partial d_j} \Delta d_j. \tag{6.55}$$

If we assume that the individual errors $\varepsilon_{i,j}$ are independent and follow the L_1 distribution of Eq. (6.4), then the total error ε_i in the ith component will be

$$\varepsilon_i = \sum_{j=1}^{m} |\varepsilon_{i,j}| = \sum_{j=1}^{m} \left| \frac{\partial X_i}{\partial d_j} \Delta d_j \right|. \tag{6.56}$$

For the error Δd_j take a constant quantity, the MAD. We arrive at

$$\varepsilon_i = \text{MAD} \sum_{j=1}^{m} \left| \frac{\partial X_i}{\partial d_j} \right|. \tag{6.57}$$

To determine the partial derivatives $\partial X_i / \partial d_j$ use Eq. (5.15). Rather than use the vector Eq. (5.15), in component form, define $\boldsymbol{\varepsilon}^T = (\varepsilon_1 \ \varepsilon_2 \cdots \ \varepsilon_m)$ and Eq. (6.57) becomes

$$\boldsymbol{\varepsilon} = \text{MAD} \sum_{j=1}^{m} |(\mathbf{A}^T \cdot \mathbf{A})^{-1} \cdot \mathbf{A}_j^T|, \tag{6.58}$$

where $\mathbf{A}_j^T$ is the jth column of $\mathbf{A}^T$. Equation (6.58) is the final equation for the error estimates.

Equation (6.58) involves fewer arithmetic operations than Monte Carlo simulation. To form $\mathbf{A}^T \cdot \mathbf{A}$ requires $mn^2/2$ operations, to calculate the inverse $(\mathbf{A}^T \cdot \mathbf{A})^{-1}$ $n^3/2$ operations, and to compute the product of the inverse with $\mathbf{A}_j^T$ a further mn operations. For large m the $mn^2/2$ term dominates; for $m \approx n$ the count will be approximately n^3. A Monte Carlo simulation using ten individual solutions would generate an operation count for approximately $100mn^2$, substantially higher, particularly when $m \gg n$. Nor does Eq. (6.58) entail having to work with the entire matrix $\mathbf{A}$. The product $\mathbf{A}^T \cdot \mathbf{A}$ may be accumulated sequentially as with the method of least squares; the columns $\mathbf{A}_j^T$ may be handled one at a time.

Equation (6.58), however, suffers a drawback not found in the Monte Carlo simulation of L_1 errors. Outliers enter into the calculation of the MAD, although with less weight than in the calculation of the mean error of unit weight, $\sigma(1)$, and, unless eliminated, the ε_i's will be unrealistic. This is an unavoidable consequence of the use of a measure of dispersion to compute errors: large residuals influence a dispersion regardless of the robustness of the solution. A Monte Carlo simulation of L_1 errors is not bothered by outliers because the individual solutions are robust.

Although not immediately evident from an inspection of Eq. (6.58), L_1 errors turn out to be larger than least squares errors. This is reasonable, given that Eq. (6.4) goes to zero far more slowly than the normal distribution and regards a large residual, which least squares would consider discordant, as normal or only slightly pernicious. It is an instructive exercise to examine the scatter in the solution vectors from Monte Carlo simulation with the L_1 distribution and the normal distribution. The latter is markedly more concentrated.

Equation (6.58) calculates errors comparable to errors found by Monte Carlo simulation but systematically smaller, a consequence of using the pseudoinverse in Eq. (5.15) to link the vectors X and d. The pseudoinverse arises naturally in least squares problems because it affords a minimum length solution in the Euclidean norm to Eq. (4.1). But a minimum length in the L_1 norm is not a minimum length in the Euclidean norm and vice versa, although the lengths do not differ much in magnitude. Then why bother with the pseudoinverse at all? Because there is no matrix that links X to d and satisfies the L_1 criterion. If such a matrix existed we could solve Eq. (6.1) by something analogous to Eq. (5.15) and not resort to the more complicated linear programming approach. The pseudoinverse meets the requirement of computational efficiency, and the bias introduced is not severe.

We finish this section with a concrete example, the L_1 solution with error estimates of Eq. (5.64), the same linear system used in Section 5.5 to illustrate iteratively reweighted least squares. Barrodale and Roberts also use this system in the explanation of their L_1 algorithm. For comparison's sake a least squares solution of this system is advisable. By the formulas of Chapter 5 we calculate:

$$X^T = (0.60000 \quad 0.40000),$$
$$\varepsilon^T = (0.66332 \quad 0.20000),$$
$$r^T = (0.00000 \quad 0.40000 \quad -0.20000 \quad -0.80000 \quad 0.60000),$$
$$\sigma(1) = 0.63246.$$

The Barrodale and Roberts algorithm calculates the solution as

$$X^T = (0.50000 \quad 0.50000)$$

with corresponding residuals

$$r^T = (0.00000 \quad 0.50000 \quad 0.00000 \quad -0.50000 \quad 1.00000).$$

In other words, the L_1 solution comes from solving the first and the third equations.

The solution is degenerate because

$$X^T = (0.75000 \quad 0.25000)$$

results in residuals

$$r^T = (0.00000 \quad 0.25000 \quad -0.50000 \quad -1.25000 \quad 0.00000)$$

with the same sum of absolute values. The Barrodale and Roberts algorithm, nevertheless, has no difficulty in obtaining a solution.

To calculate the L_1 errors we have from Eq. (6.54)

$$\text{MAD} = \frac{0 + 0.5 + 0 + 0.5 + 1}{5 - 2} = 0.66667,$$

and

$$A^T \cdot A = \begin{pmatrix} 5 & 15 \\ 15 & 55 \end{pmatrix},$$

from which

$$(\mathbf{A}^T \cdot \mathbf{A})^{-1} = \begin{pmatrix} 1.1 & -0.3 \\ -0.3 & 0.1 \end{pmatrix}.$$

From Eq. (6.58)

$$\begin{pmatrix} \varepsilon_1 \\ \varepsilon_2 \end{pmatrix} = 0.66667 \begin{cases} |0.8| + \ |0.5| + |0.2| + |-0.1| + |-0.4| \\ |-0.2| + |-0.1| + |0| \ + | \ 0.1| + | \ 0.2|, \end{cases}$$

or

$$\varepsilon^T = (1.33333 \quad 0.40000).$$

To summarize, the L_1 solution results in:

$$\mathbf{X}^T = (0.50000 \quad 0.50000),$$
$$\varepsilon^T = (1.33333 \quad 0.40000),$$
$$\mathbf{r}^T = (0.00000 \quad 0.50000 \quad 0.00000 \quad -0.50000 \quad 1.00000),$$
$$\text{MAD} = 0.66667.$$

As stated, the L_1 errors are larger than the least squares errors despite the MAD and $\sigma(1)$ being nearly the same; the MAD is only 5% larger.

```
C
C
C    THIS PROGRAM DETERMINES THE SOLUTION OF AN OVERDETERMINED SYSTEM IN THE L1 NORM.
C
C
           PROGRAM L1
           IMPLICIT REAL *8 ( A - H , O - Z )
           REAL *8  MIN , MAX
           REAL *16  SUM
           INTEGER *4  OUT, S ( 1000 )
           LOGICAL STAGE, TEST
           DIMENSION  A ( 1000 , 1000 ) , X ( 1000 ) , E ( 1000 ) , B ( 1000 )
           DATA  BIG / 1.0D38 /
           TOLER = 1 D - 10
           TYPE *, 'WHAT IS M ? '
           ACCEPT *, M
           TYPE *, 'WHAT IS N? '
           ACCEPT *, N
           INDEX = 0
           DO  I = 1 , M
             DO  J = 1 , N
               WRITE ( *, 5) I , J
5              FORMAT ( ' A ( ',I2, ', ',I2, ') = ? ')
               READ ( *, 10) A ( I , J )
10             FORMAT ( D22.15 )
             END  DO
           END  DO
           TYPE *, 'WHAT IS THE RIGHT-HAND SIDE ? '
             DO  I = 1 , M
```

Figure 6.4. L_1 Solution of overdetermined system.

```
                    WRITE ( * , 15 ) I
15                  FORMAT ( ' B ( ' , I 2 , ' ) = ? ' )
                    READ ( * , 10 ) B ( I )
               END DO
C
C     INITIALIZATION
C
                    DO J = 1 , N
                      A ( M + 2 , J ) = J
                      X ( J ) = 0D0
                    END DO
                    DO I = 1 , M
                      A ( I , N + 2 ) = N + I
                      A ( I , N + 1 ) = B ( I )
                      IF ( B ( I ) . LT . 0D0 ) THEN
                        DO J = 1 , N + 2
                          A ( I , J ) = - A ( I , J )
                        END DO
                      END IF
                      E ( I ) = 0D0
                    END DO
C
C     COMPUTE THE MARGINAL COSTS
C
                    DO J = 1 , ( N + 1 )
                      SUM = 0.0
                      DO I = 1 , M
                         SUM = SUM + A ( I , J )
                      END DO
                      A ( M + 1 , J ) = SUM
                    END DO
C
C
C               STAGE I
C
C
C
C     DETERMINE THE VECTOR TO ENTER THE BASIS
C
                    STAGE = .TRUE.
                    KOUNT = 0
                    KR = 1
                    KL = 1
                    DO WHILE ( KOUNT + KR .NE. N + 1 )
                      MAX = - 1 D0
                      DO J = KR , N
                        IF ( ABS ( A ( M + 2 , J ) ) .LE. N ) THEN
                          D = ABS ( A ( M + 1 , J ) )
                            IF ( D .GT. MAX ) THEN
                              MAX = D
                              IN = J
                            END IF
                        END IF
                      END DO
                      IF ( A ( M + 1 , IN ) .LT. 0D0 ) THEN
                        DO I = 1 , M + 2
                        A ( I , IN ) = - A ( I , IN )
```

Figure 6.4 (*continued*)

```
                        END  DO
                        END  IF
C
C     DETERMINE THE VECTOR TO LEAVE THE BASIS
C
100                     K = 0
                        DO  I = KL , M
                          D = A ( I , IN )
                          IF ( D .GT. TOLER ) THEN
                            K = K + 1
                            B ( K ) = A ( I , N + 1 ) / D
                            S ( K ) = I
                            TEST = .TRUE.
                          END  IF
                        END  DO
120                     IF ( K .LE. 0 ) THEN
                          TEST = .FALSE.
                        ELSE
                          MIN = BIG
                          DO  I = 1 , K
                            IF ( B ( I ) .LT. MIN ) THEN
                              J = I
                              MIN = B ( I )
                              OUT = S ( I )
                            END  IF
                          END  DO
                          B ( J ) = B ( K )
                          S ( J ) = S ( K )
                          K = K - 1
                        END  IF
C
C     CHECK FOR LINEAR DEPENDENCE IN STAGE I
C
                        IF ( . NOT . TEST . AND . STAGE ) THEN
                          DO  I = 1 , M + 2
                            D = A ( I , KR )
                            A ( I , KR ) = A ( I , IN )
                            A ( I , IN ) = D
                          END  DO
                          KR = KR + 1
                        ELSE
                          IF ( . NOT . TEST ) THEN
                            A ( M + 2 , N + 1 ) = 2D0 )
                            GOTO  350
                          END  IF
                          PIVOT = A ( OUT , IN )
                          IF ( A ( M + 1 , IN ) - 2.0 * PIVOT . GT . TOLER ) THEN
                            DO  J = KR , N + 1
                              D = A ( OUT , J )
                              A ( M + 1 , J ) = A ( M + 1 , J ) - 2.0 * D
                              A ( OUT , J ) = - D
                            END  DO
                            A ( OUT , N + 2 ) = - A ( OUT , N + 2 )
                            GOTO  120
                          END  IF
C
C     PIVOT ON  A ( OUT , IN )
```

Figure 6.4 (*continued*)

```
C
                    DO J = KR,N + 1
                      IF (J.NE.IN) A(OUT,J) = A(OUT,J)/PIVOT
                    END DO
                    DO I = 1,M + 1
                      IF (I.NE.OUT) THEN
                        D = A(I,IN)
                        DO J = KR,N + 1
                          IF (J.NE.IN) A(I,J) = A(I,J)-D * A(OUT,J)
                        END DO
                      END IF
                    END DO
                    DO I = 1,M + 1
                      IF (I.NE.OUT) A(I,IN) = -A(I,IN)/PIVOT
                    END DO
                    A(OUT,IN) = 1D0/PIVOT
                    D = A(OUT,N + 2)
                    A(OUT,N + 2) = A(M + 2,IN)
                    A(M + 2,IN) = D
                    KOUNT = KOUNT + 1
                    IF (.NOT.STAGE) GOTO 270
C
C    INTERCHANGE ROWS IN STAGE I
C
                    KL = KL + 1
                    DO J = KR,N + 2
                      D = A(OUT,J)
                      A(OUT,J) = A(KOUNT,J)
                      A(KOUNT,J) = D
                    END DO
                  END IF
                END DO
C
C
C          STAGE II
C
C
                STAGE = .FALSE.
C
C    DETERMINE THE VECTOR TO LEAVE THE BASIS
C
270                MAX = -BIG
                DO J = KR,N
                  D = A(M + 1,J)
                  IF (D.GE.0D0) THEN
                    IF (D.GT.MAX) THEN
                      MAX = D
                      IN = J
                    END IF
                  ELSE
                    IF (D.LE.(-2D0)) D = -D-2D0
                  END IF
                END DO
                IF (MAX.GT.TOLER) THEN
                  IF (A(M + 1,IN).GT.0D0) GOTO 100
                  DO I = 1,M + 2
                    A(I,IN) = -A(I,IN)
```

Figure 6.4 (*continued*)

```
                END DO
                A(M + 1, IN) = A(M + 1, IN) - 2D0
                GOTO 100
             END IF
C
C    PREPARE OUTPUT
C
             L = KL - 1
             DO I = 1, L
               IF (A(I, N + 1). LT . 0D0) THEN
                  DO J = KR, N + 2
                     A(I, J) = -A(I, J)
                  END DO
               END IF
             END DO
             A(M + 2, N + 1) = 0D0
             IF (KR. EQ. 1) THEN
                DO J = 1, N
                   D = ABS(A(M + 1, J))
                   IF (D. LE . TOLER . OR . 2D0 - D . LE . TOLER) GOTO 350
                END DO
                A(M + 2, N + 1) = 1D0
             END IF
350          DO I = 1, M
               K = A(I, N + 2)
               D = A(I, N + 1)
               IF (K. LE . 0) THEN
                  K = -K
                  D = -D
               END IF
               IF (I. GE . KL) THEN
                  K = K - N
                  E(K) = D
               ELSE
                  X(K) = D
               END IF
             END DO
             A(M + 2, N + 2) = KOUNT
             A(M + 1, N + 2) = N + 1 - KR
             SUM = 0.0
             DO I = KL, M
               SUM = SUM + A(I, N + 1)
             END DO
             A(M + 1, N + 1) = SUM
             TYPE * , ' THE SOLUTION VECTOR IS : '
             WRITE( * , 105) (X(I), I = 1, N)
105          FORMAT ( ' ', D22.15, '
             TYPE * , '
             TYPE * , ' THE RESIDUALS ARE : '
             WRITE( * , 105) (E(I), I = 1, M)
             TYPE * , '
             TYPE * , ' ', SUM, ' IS THE SUM OF THE ABSOLUTE VALUES OF THE RESIDUALS '
             TYPE * , '
             TYPE * , ' ', KOUNT, ' ITERATIONS WERE PERFORMED
             TYPE * , '
             TYPE * , ' ', N + 1 - KR, ' IS THE RANK OF THE MATRIX
             END
```

Figure 6.4 (*continued*)

References

Abdelmalek, N.N. (1971). Linear L_1 Approximation for a Discrete Point Set and L_1 Solutions of Overdetermined Linear Equations, *J. ACM*, **18**, p. 41.

Abdelmalek, N.N. (1975). An Efficient Method for the Discrete Linear L_1 Approximation Problem, *Math. Comp.*, **29**, p. 844.

Armstrong, R.D. and Godfrey, J.P. (1979). Two Linear Programming Algorithms for the Linear Discrete L_1 Norm Problem, *Math. Comp.*, **33**, p. 289.

Barrodale, I. and Roberts, F.D.K. (1973). An Improved Algorithm for Discrete L_1 Linear Approximation, *SIAM J. Numer. Anal.*, **10**, p. 839.

Barrodale, I. and Roberts, F.D.K. (1974). Algorithm 478: Solution of an Overdetermined System of Equations in the L_1 Norm, *Commun. ACM*, **17**, p. 319.

Bartels, R.H., Conn., A.R., and Sinclair, J.W. (1978). Minimization Techniques for Piecewise Differentiable Functions: The L_1 Solution to an Overdetermined Linear System, *SIAM J. Numer. Anal.*, **15**, p. 224.

Branham, R.L., Jr. (1986). Error Estimates with L_1 Solutions, *Celest. Mech.*, **39**, p. 239.

Forsythe, G., Malcolm, M.A., and Moler, C.B. (1977). *Computer Methods for Mathematical Computations* (Prentice-Hall, Englewood Cliffs, N.J.).

Franklin, J. (1987). Convergence in Karmarkar's Algorithm for Linear Programming, *SIAM J. Num. Anal.*, **24**, p. 928.

Gass, S.I. (1975). *Linear Programming: Methods and Applications* (McGraw-Hill, New York).

Gauss, K.F. (1963). *Theory of the Motion of the Heavenly Bodies Moving about the Sun in Conic Sections* (Dover, New York).

Karmarkar, N. (1984). A New Polynomial–Time Algorithm for Linear Programming, *Combinatorica*, **4**, p. 373.

Patterson, E.M. (1969). *Topology* (Interscience, New York).

Press, W.H., Flannery, B.P., Teukolsky, S.A., and Vetterling, W.T. (1986). *Numerical Recipes: The Art of Scientific Computing* (Cambridge University Press, Cambridge).

Tewarson, R.P. (1973). *Sparse Matrices* (Academic Press, Orlando, Florida).

Zelen, M. and Severo, N.C. (1965). *Probability Functions*. In Abramowitz, M. and Stegun, I.A. (eds.) *Handbook of Mathematical Functions* (Dover, New York).

CHAPTER 7

Nonlinear Methods

7.1. Introduction

Sometimes our problem involves nonlinear equations. If, for example, we wish to determine the half-life of a radioactive nuclide. In the p-norm the equations assume the form

$$\|A_0 e^{-\lambda t_i} - A_i\|_p = \min, \tag{7.1}$$

where A_0 is the original quantity of nuclide at time $t_0 = 0$, A_i is the quantity at time t_i, and λ is the half-life to be determined. Physical chemistry offers the Michaelis–Menten equation: X_i is the concentration of an enzyme with a reaction rate y_i and we wish to determine constants a and b such that

$$\left\| \frac{aX_i}{b + X_i} - y_i \right\|_p = \min. \tag{7.2}$$

In general terms, we minimize the p-norm of a nonlinear function $h(\mathbf{X})$,

$$\|h(\mathbf{X}) - \mathbf{d}\|_p = \min, \tag{7.3}$$

where $h(\mathbf{X})$ may no longer be written as $\mathbf{A} \cdot \mathbf{X}$ as in Eq. (4.1). Usually, but not always, our interest is $p = 2$, the criterion of least squares.

Occasionally, we may find a transformation that reduces a nonlinear equation to a linear one. With Eq. (7.1) and the least squares criterion, for example, the equation reduces to

$$\sum_{i=1}^{m} (\ln A_0 - \lambda t_i - \ln A_i)^2 = \min \tag{7.4}$$

upon our taking natural logarithms of both sides of the equation. The Michaelis–Menten equation may be written as

$$\sum_{i=1}^{m} \left(\frac{b}{a} \frac{1}{X_i} + \frac{1}{a} - \frac{1}{y_i} \right)^2 = \min. \tag{7.5}$$

By defining new variables $X_i' = 1/X_i$ and $y_i' = 1/y_i$ we obtain a linear equation for a and b.

But these transformations also bias the results. Equation (7.4) minimizes the sum of the squares of the logarithms of the residuals, not the residuals themselves. Likewise, with the Michaelis–Menten equation the actual residuals minimized bear a complicated relationship with the residuals we wish to minimize. The parameters—λ in Eq. (7.4) and a and b in Eq. (7.5)—calculated from a linear transformation will not be the same as those obtained from a solution of the original nonlinear equation.

Consider a simple example of curve fitting. Suppose we wish to pass the curve

$$y = ae^{bx} \tag{7.6}$$

through the pairs of X, y points (0, 1), (1, 2), and (2, 6) using the criterion of least squares. The logarithmic transformation

$$\ln y = \ln a + bX \tag{7.7}$$

results in $a = 0.93466$ and $b = 0.89588$. One of the nonlinear methods discussed in the next section, the Gauss–Newton iteration, gives $a = 0.78044$ and $b = 1.01742$. As expected the latter parameter values yield a smaller sum square of the residuals, 0.07423, than the former, 0.24174. Therefore, even if a transformation to convert a nonlinear equation to a linear equation is available, we should probably not use it. We should only use the logarithmic transformation incorporated in Eq. (7.4), if our genuine interest is to minimize the sum of the squares of the logarithms of the residuals. To handle nonlinear equations properly we need nonlinear parameter fitting techniques.

We should distinguish carefully between genuine nonlinear equations and equations that, while employing nonlinear functions, are actually linear. Equation (7.1) is nonlinear because the parameter to be determined occurs in the exponential. The nonlinear function $h(X)$ cannot be expressed as $\mathbf{A} \cdot \mathbf{X}$. A companion problem, given a nuclide with known half-life λ, measured quantities A_i at time t_i, and determining the quantity A_0 originally present at time $t = 0$, is actually linear: the parameter of interest, A_0, is not found in the exponential. The equation may be recast in the form of Eq. (4.1) with $\mathbf{X}$ a 1×1 vector of the unknown A_0, $\mathbf{A}$ the $m \times 1$ vector $(e^{\lambda t_1} \ e^{\lambda t_2} \ \ldots \ e^{\lambda t_m})$, and $\mathbf{d}$ the $m \times 1$ vector $(A_1 \ A_2 \ \ldots \ A_m)$.

Many techniques for nonlinear parameter estimation have been developed, and it is an area undergoing continuous research. In this chapter we cannot even begin to present a representative survey. Instead the author will concentrate on a few methods with which he has had practical experience. Kennedy and Gentle (1980, Chap. 10) provide a relatively complete survey of the principal nonlinear methods. The methods given in this chapter are frequently used, although none of them can be considered standard in the sense that Cholesky decomposition is the standard method for solving normal equations because a standard nonlinear method does not exist. Nonlinear problems are considerably more complicated than linear ones, and a technique that works well for one problem may fail for another. We should, therefore, maintain a library of several methods to use when tackling nonlinear problems.

Nonlinear methods fall into two general categories: those that use the gradient; and those that do not. Most readers are presumably familiar with the gradient, denoted in this book as either $\mathbf{g}$ or $\mathbf{g(X)}$. In a Cartesian coordinate system the gradient at point $\mathbf{X}$ in the n space of a function $f(\mathbf{X})$ is simply, as used in Section 6.3,

$$\mathbf{g}^{\mathrm{T}}(\mathbf{X}) = \left(\frac{\partial f}{\partial X_1} \quad \frac{\partial f}{\partial X_2} \quad \cdots \quad \frac{\partial f}{\partial X_n} \right). \tag{7.8}$$

Nonlinear methods are by nature iterative. We need an initial approximation to the solution and hope to refine this approximation, but we cannot say how many iterations will be required to refine the solution to a specified precision. Indeed, the iterations may not even converge. Gradient methods are generally more efficient than nongradient methods, in terms of the number of iterations required, because the gradient supplies the valuable information of the direction of maximum increase of the function. But nongradient methods are more robust and converge when gradient methods diverge.

Before starting the discussion of the methods we should clarify two matters. We talk about the convergence of the solution, but there are numerous ways to stipulate a convergence criterion. Perhaps the easiest way monitors the components of the solution vector at iteration $n + 1$ compared with the components at a previous iteration, say the nth, and accepts convergence when the differences between the components are less than an established tolerance. But this convergence criterion must be applied with care because some methods make little progress for a number of iterations and then calculate a large correction to the current solution. Once the differences in the components from the previous and the current iterations fall below the tolerance, it is better to let the program grind away for a few more iterations to assure that the corrections remain below the tolerance, indicating that convergence has indeed been achieved. We may, in addition or instead, monitor the components of the gradient, when using a gradient method, and the differences in the calculated minimum of Eq. (7.1),

$$F(\mathbf{X}) = \sum_{i=1}^{m} |h_i(\mathbf{X}) - d_i|^p. \tag{7.9}$$

The second matter concerns terminology. When $p = 2$ it is customary to refer to the nonlinear method as a problem of nonlinear regression, and when p is other than 2 as a problem of nonlinear optimization. Methods for nonlinear optimization may also be applied to the case $p = 2$, but tend to be less efficient because when $p = 2$ the function $F(\mathbf{X})$ assumes a special form that can be exploited, as we see in the next section.

7.2. Gradient Methods

Methods that use the gradient need an initial approximation to the solution vector to start the iterations. Unless the initial approximation is close, in some sense of the term, to the true solution, the method may diverge; or it may converge to a local, rather than a global, minimum. The topology of the function (7.9) can be extremely complicated, with many local minima. Although it is possible to give strict mathematical definitions to the terms "close" and "local minimum", such definitions are useless for practical purposes. In the next section we present a method for obtaining an initial approximation that should be adequate for most problems.

Numerous algorithms for the nonlinear optimization of a function are available. Before discussing one of the best of these, the Davidon–Fletcher–Powell (DFP) algorithm, we start with a simpler algorithm, called steepest descent. Cauchy first proposed the steepest descent method in 1847 in connection with solving systems of nonlinear equations. Although too inefficient for general use, it serves as a convenient starting point for the discussion of better algorithms.

Because the gradient indicates the direction of the maximum increase of a function, if we calculate the gradient $g(\mathbf{X})$ at point $\mathbf{X}$ and take a small step in the opposite direction the function must decrease. Cauchy suggested a move in the direction

$$\mathbf{X}_{n+1} = \mathbf{X}_n - \alpha g(\mathbf{X}_n) \qquad (0 < \alpha < \infty) \tag{7.10}$$

from the point $\mathbf{X}_n$ to the point $\mathbf{X}_{n+1}$ to decrease the function, $f(\mathbf{X}_{n+1}) < f(\mathbf{X}_n)$. The parameter α may be determined in a number of ways. We may take α as a small constant, such as 0.001, use it, and hope for the best. But such a simple-minded procedure will likely turn the inefficient steepest descent method even more inefficient; we would like α to be as large as possible.

Suppose that we are not too far from the minimum, $\mathbf{X}_{min}$; we are actually at point $\mathbf{X}_0$, a nearby point. From elementary calculus

$$df(\mathbf{X}_{min}) = \frac{\partial f(\mathbf{X}_{min})}{\partial \mathbf{X}_{min}} \cdot d\mathbf{X}_{min}. \tag{7.11}$$

We reach the minimum from point $\mathbf{X}_0$ by setting Eq. (7.11) equal to $-f(\mathbf{X}_0)$,

$$\frac{\partial f(\mathbf{X}_{min})}{\partial \mathbf{X}_{min}} \cdot d\mathbf{X}_{min} = -f(\mathbf{X}_0). \tag{7.12}$$

Replacing $d\mathbf{X}_{min}$ by $\Delta \mathbf{X} = \mathbf{X}_{min} - \mathbf{X}_0$, using Eq. (7.10) with $\mathbf{X}_{min} = \mathbf{X}_{n+1}$ and $\mathbf{X}_0 = \mathbf{X}_n$, and remembering that $\mathbf{X}_0$ is close to $\mathbf{X}_{min}$ and, therefore, $\mathbf{X}_{min} \cong \mathbf{X}_0$, Eq. (7.12) becomes

$$\alpha = f(\mathbf{X}_0) \left/ \frac{\partial f^{\mathrm{T}}}{\partial \mathbf{X}_0} \cdot \frac{\partial f}{\partial \mathbf{X}_0} \right., \tag{7.13}$$

with the derivative of the function evaluated at $\mathbf{X}_0$.

Other, and better, possibilities for determining α exist, as we shall see in a moment. Equation (7.13) depends on our being close to the minimum, an assumption of relatively dubious validity, for the first iterations at least. But Eq. (7.13) at any rate relies on a modicum of mathematical theory, unlike arbitrarily choosing some constant value for α.

The steepest descent method converges more slowly, when it converges, than other gradient methods. Part of the problem resides in having to move in the direction of the new gradient, perpendicular to the previous direction, at each iteration instead of proceeding via the shortest route towards the function minimum. For certain geometries, such as highly elongated ellipses in two-dimensional space, we move in a series of short zigzags, as Figure 7.1 shows.

To illustrate the slow convergence of the steepest descent method, consider finding the solution of the equation

$$f(X_1, X_2) = (X_1 - 2)^2 + 10{,}000(X_2 - 1)^2, \qquad (7.14)$$

which obviously is $X^T = (2\ 1)$. Starting with a first approximation of $X_0^T = (0\ 0)$, working to six decimal digit precision, or a tolerance of $5 \cdot 10^{-7}$, and using Eq. (7.13) to calculate α, the steepest descent method needs 1,930 iterations to calculate the solution.

Despite its slowness, the steepest descent method is easy to program. Some thirty or forty lines of code suffice; even that worst of programming languages for numerical computations, BASIC, can be used. The method may, therefore, be useful on home computers, where the user merely starts his steepest descent algorithm and lets it grind away to a solution without worrying about the excessive time consumed because the computer has nothing else to do anyway nor is the user paying for CPU time.

But for efficiency some way is needed to accelerate the convergence. The DFP method, mentioned previously, is one of the best of what are known as quasi-Newton or variable metric methods for nonlinear optimization. Given our function of n variables $f(X)$, expand it about the minimum point X_{min} by a Taylor series up to second-order terms:

$$f(X) = f(X_{min}) + (X - X_{min})^T \cdot g(X_{min}) + \tfrac{1}{2}(X - X_{min})^T \cdot H \cdot (X \cdot X_{min}), \quad (7.15)$$

where H is the symmetric Hessian matrix of the second-order partial derivatives of the function evaluated at X_{min},

$$H(X_{min}) = \begin{pmatrix} \dfrac{\partial^2 f(X_{min})}{\partial^2 X_{min,1}} & \dfrac{\partial^2 f(X_{min})}{\partial X_{min,1}\,\partial X_{min,2}} & \cdots & \dfrac{\partial^2 f(X_{min})}{\partial X_{min,1}\,\partial X_{min,n}} \\[2ex] \dfrac{\partial^2 f(X_{min})}{\partial X_{min,2}\,\partial X_{min,1}} & \dfrac{\partial^2 f(X_{min})}{\partial^2 X_{min,2}} & \cdots & \dfrac{\partial^2 f(X_{min})}{\partial X_{min,2}\,\partial X_{min,n}} \\[2ex] \vdots & & & \\[2ex] \dfrac{\partial^2 f(X_{min})}{\partial X_{min,n}\,\partial X_{min,1}} & \dfrac{\partial^2 f(X_{min})}{\partial X_{min,n}\,\partial X_{min,2}} & \cdots & \dfrac{\partial^2 f(X_{min})}{\partial^2 X_{min,n}} \end{pmatrix}.$$

$$(7.16)$$

Because the gradient vanishes at the minimum the second term on the right-hand side of Eq. (7.15) is zero. The gradient at X, found from differentiation

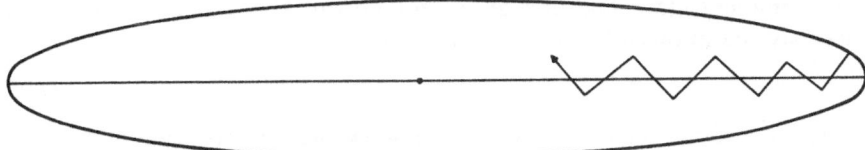

Figure 7.1. Steepest descent zigzags towards minimum in highly elliptical geometry.

of Eq. (7.15) is

$$g(X) = H(X_{min}) \cdot (X - X_{min}). \tag{7.17}$$

To find the minimum X_{min}, therefore, we set

$$X_{min} = X - H^{-1}(X_{min})g(X). \tag{7.18}$$

Equation (7.18) shows one difficulty of the steepest descent method. Consider the function (7.14). Its gradient is

$$g(X) = \begin{pmatrix} 2(X_1 - 2) \\ 20,000(X_2 - 1) \end{pmatrix}. \tag{7.19}$$

Evaluated at $X^T = (0\ 0)$ (7.19) is $g^T(X) = (-4\ -20,000)$. With α determined from Eq. (7.13), the steepest descent method takes a step of size $2.5 \cdot 10^{-5}$ in the direction of $g^T(X)$. But Eq. (7.18) shows that we should really move in the direction

$$-H^{-1} \cdot g = -\begin{pmatrix} 0.5 & 0 \\ 0 & 0.5 \cdot 10^{-4} \end{pmatrix} \begin{pmatrix} -4 \\ 2 \cdot 10^4 \end{pmatrix} = \begin{pmatrix} 2 \\ 1 \end{pmatrix}. \tag{7.20}$$

One step takes us to the minimum with Eq. (7.20); but the gradient approximates extremely poorly the direction towards the minimum.

Most functions, of course, are not nice quadratics like (7.14) but we nevertheless expect that near the minimum a properly behaving function should be well approximated by Eq. (7.15). The DFP method, therefore, proposes that instead of Eq. (7.10) we use

$$X_{i+1} = X_i - \alpha H^{-1} \cdot g, \tag{7.21}$$

with H^{-1} and g evaluated at X_i, to move towards the minimum. Calculating the inverse Hessian matrix with each iterative step is unappealing, and the DFP method uses an approximation, easier to calculate, to H^{-1}. We will sketch the arguments used to approximate H^{-1}. Kennedy and Gentle (1980, Sec. 10.2.3) give more details.

The Hessian matrix is symmetric and near the minimum positive definite. These properties should be preserved in the approximation to H^{-1}. (Most sources use the notation H instead of H^{-1}, but for an inverse Hessian matrix such notation seems aberrant.) Furthermore, the job should be done using the information at hand: successive values of X, of the function $f(X)$, and of the gradient $g(\mathbf{X})$. The first approximation to $\mathbf{H}^{-1}$ is the unit matrix $\mathbf{I}$. Successive iterations add a correction term $\mathbf{c}_i$

$$\mathbf{H}_{i+1}^{-1} = \mathbf{H}_i^{-1} + \mathbf{c}_i, \tag{7.22}$$

where $\mathbf{c}_i$ conserves the properties of symmetry and positive definiteness.

Start out with a vector

$$\mathbf{s}_i = -\mathbf{H}_i^{-1} \cdot \mathbf{g}_i, \tag{7.23}$$

a step in the direction of the negative gradient modified by $\mathbf{H}^{-1}$. For the first iteration $\mathbf{H}^{-1}$ is $\mathbf{I}$ and we move in the direction of the gradient. How far do we move in this direction? A distance α. To find α we could use Eq. (7.13), but for an efficient algorithm something more sophisticated is needed, which we give shortly. Calculate a vector

$$\mathbf{d}_i = \alpha \mathbf{s}_i \tag{7.24}$$

and modify our current estimate of the minimum $\mathbf{X}_i$ by $\mathbf{d}_i$,

$$\mathbf{X}_{i+1} = \mathbf{X}_i + \mathbf{d}_i. \tag{7.25}$$

Calculate the difference in the gradient

$$\mathbf{y}_i = \mathbf{g}_{i+1} - \mathbf{g}_i. \tag{7.26}$$

In 1959, Davidon proved that the correction term $\mathbf{c}_i$ can be expressed as the sum of two matrices:

$$\mathbf{A}_i = \frac{\mathbf{d}_i \cdot \mathbf{d}_i^{\mathsf{T}}}{\mathbf{d}_i^{\mathsf{T}} \cdot \mathbf{y}_i} \tag{7.27}$$

and

$$\mathbf{B}_i = \frac{(\mathbf{H}_i^{-1} \cdot \mathbf{y}_i)^{\mathsf{T}} \cdot (\mathbf{H}_i^{-1} \cdot \mathbf{y}_i)}{\mathbf{y}_i^{\mathsf{T}} \cdot \mathbf{H}^{-1} \cdot \mathbf{y}_i}. \tag{7.28}$$

In both Eqns. (7.27) and (7.28) the numerator is the backwards product of two vectors and the denominator the scalar product. Each matrix, therefore, is of rank one and their sum of rank two. Kennedy and Gentle (1980, Sec. 10.2.3) prove that $\mathbf{c}_i$ is positive definite—from Eqns. (7.27) and (7.28) it is obviously symmetric—and that successive iterates $\mathbf{H}_i^{-1} + \mathbf{c}_i$ converge to the inverse Hessian matrix and converge quadratically.

What does "quadratic convergence" mean? Let the true solution be $\mathbf{X}_{\min}$ and the current approximation to the solution be $\mathbf{X}_i$. The error of the approximation is $\mathbf{e}_i = \mathbf{X}_i - \mathbf{X}_{\min}$. By the Forsythe, Malcolm, and Moler (1977, p. 158) definition a convergent iteration that satisfies

$$\lim_{h \to \infty} \frac{\|\mathbf{e}_h + 1\|}{\|\mathbf{e}_h^d\|} = \text{constant} \tag{7.29}$$

has dth order convergence. When $d = 2$ the convergence is said to be quadratic. This means that roughly we pick up a squared number of digits of precision with each iteration.

The matrices $\mathbf{A}_i$ and $\mathbf{B}_i$ are not the only way to form the undate matrix $\mathbf{c}_i$.

Another, and somewhat more accurate, update is known as the Broyden–Fletcher–Greenstadt–Goldfarb–Shanno correction matrix, discussed in Kennedy and Gentle (1980, Sec. 10.2.3).

What about α? Equation (7.13) is only a good approximation near the minimum. More refined techniques either search in a given direction until a minimum of the function is found or calculate a few function values, or function and gradient values, and pass a smooth curve such as a parabola or cubic through the points. After differentiation the minimum may be calculated immediately.

To use the first technique we need to bracket an interval where the function is a minimum. Although bracketing is not strictly required for the second technique, it is, nevertheless, a good idea. Unless we know that the function has a minimum in a given interval a formally calculated minimum may be more fictious than real. To bracket an interval with a minimum we take an arbitrary α, calculate $f(0)$, $f(\alpha)$, $f(2\alpha)$, $f(2^2\alpha)$, and keep doubling α until $f(2^n\alpha) > f(0)$.

With the function (7.14), for example, if we start at $X^T = (0 \ 0)$ and search in the direction of the gradient (7.19) using a step size of $\alpha = 10^{-5}$, a value suggested by Eq. (7.13)—which gives $2.5 \cdot 10^{-5}$—as an appropriate scale for the problem, we find

α	$f(\alpha)$
0	10004
$1 \cdot 10^{-5}$	6404
$2 \cdot 10^{-5}$	3604
$4 \cdot 10^{-5}$	404
$8 \cdot 10^{-5}$	3604

The bracketing interval, therefore, is $[0, 8 \cdot 10^{-5}]$. For a quick selection of α we could simply take the value that gives the smallest $f(\alpha)$, here $\alpha = 4 \cdot 10^{-5}$. But better techniques for finding α are available.

One of these, known as the Fibonacci search, can be shown to be optimal, where optimal means determining the smallest interval bracketing the minimum for a given number of function evaluations (Forsythe, Malcolm, and Moler, 1977, Sec. 8.1). The Fibonacci search employs the well-known numbers of the same name, defined by

$$F_0 = 0, \qquad F_1 = 1, \qquad F_k = F_{k-1} + F_{k-2}. \qquad (7.30)$$

To start the Fibonacci search we specify the tolerance desired. For minimization problems the tolerance should be no smaller than the square root of the machine epsilon for the precision being used. From Eq. (7.15) when the difference between X and $X_{\min}$ is of the order of the machine epsilon

$$f(X + \varepsilon) \cong f(X) + \tfrac{1}{2}\varepsilon^T \cdot H \cdot \varepsilon, \qquad (7.31)$$

where ε is an n vector whose components are ε. Thus, a change of the order of ε in $\mathbf{X}$ causes a change of the order ε^2 in $f(\mathbf{X})$. Specifying a tolerance smaller than the square root of the machine epsilon is pointless because the function values will not change within such a small interval.

Suppose that our bracketing interval is $[a, b]$ and we want to reduce the interval until it is of the order of the specified tolerance. From Eq. (7.30) the ratios $X_{k-1} = F_{k-2}/F_k$ and $X_k = F_{k-1}/F_k$ sum to unity and each X may, therefore, be interpreted as a point within the interval $[0, 1]$. We start the Fibonacci search with a given k and calculate the X's. The interval $[0, 1]$ is transformed to $[a, b]$ and we calculate the function values $f(X_k)$ and $f(X_{k-1})$. (In general the function depends on an n vector $\mathbf{X}$ rather than a scalar X, but for the moment it is easier to think of the X's as scalars.) If $f(X_k) > f(X_{k-1})$ the new bracketing interval is $[a, X_k]$. Otherwise it is $[X_{k-1}, b]$. The process continues until we reach the Fibonacci number F_1.

How do we determine the k to start the Fibonacci search? The first step of the Fibonacci search reduces the size of the bracking interval from one to F_{k-1}/F_k, the second step from F_{k-1}/F_k to F_{k-2}/F_{k-1}, and so forth up to F_1/F_2. Upon multiplying these factors together we find as a criterion for the size of the smallest interval

$$\frac{1}{F_k} < \text{toler.} \tag{7.32}$$

For single-precision on a VAX computer, for example, if we specify for the tolerance the square root of the machine epsilon, we calculate

$$\frac{1}{F_k} < 2.44 \cdot 10^{-4}.$$

From a table of the Fibonacci numbers or by calculating them we find that $F_{19} = 4{,}181$ meets the criterion. We then start the search with $k = 19$.

Figure 7.2 gives FORTRAN code for a Fibonacci search. Note that the code works with a vector function $f(\mathbf{X})$, not a scalar function $f(X)$. The function subroutine for calculating the Fibonacci numbers uses the definition (7.30). To find the kth Fibonacci number, k operations, although the simple operations of addition, are required. More efficient, but also more complicated, algorithms that calculate the kth number with $\log_2 k$ operations exist. See Dromey (1982, Sec. 3.8) for details.

The Fibonacci search always converges and converges to the minimum. Returning to the example of the function (7.14), a Fibonacci search starting from $\mathbf{X}^T = (0 \ 0)$ in the direction of the gradient calculates $\alpha = 5 \cdot 10^{-5}$, whereas Eq. (7.13) gives $\alpha = 2.5 \cdot 10^{-5}$. The former gives $f(\mathbf{X}) = 4$ and the latter $f(\mathbf{X}) = 2{,}504$. But the Fibonacci search is also slow. We can prove that

$$\lim_{k \to \infty} \frac{F_k - 1}{F_k} \to \frac{\sqrt{5} - 1}{2} = 0.61803$$

(Scheid, 1968, pp. 185–186). With each step of the search we decrease the size of the interval by 0.61803, or not quite one bit.

For this reason faster ways to find the minimum of a function in a given direction are used. One of the simplest of these fits a parabola to three function values and calculates the position of the minimum. Using once again the example of the function (7.14) suppose that we select the three pairs of values

α	$f(\alpha)$
0	10,004
$2 \cdot 10^{-5}$	3,604
$8 \cdot 10^{-5}$	3,604

The equation of a parabola is $y = A + BX + CX^2$. With these data we find $A = 10{,}004$, $B = -4 \cdot 10^8$, and $C = 4 \cdot 10^{12}$. The minimum occurs when $dy/dx = 0$, or $X_{\min} = B/2C$. For our data $X_{\min} = 5 \cdot 10^{-5}$, in good agreement with the value found from the Fibonacci search. (Good, not perfect, agreement because the Fibonacci search really gives $\alpha = 5.0000002191 \cdot 10^{-5}$.) Because the calculated $X_{\min}$ could just as easily be a maximum instead of a minimum, we should check the sign of C. For our data C is positive. The parabola is concave upward and $X_{\min}$ really is a minimum. Acton (1970, pp. 454–458) gives formulas that use both function and gradient values to fit a cubic.

Curve fitting is more rapid than a Fibonacci search for finding a minimum, but also less robust. We can always fit a parabola, which need not even straddle the minimum, to three points, but the calculated minimum may have little to do with the actual minimum of the function in that direction. Brent (1973, Chap. 5) published an algorithm that combines the virtues of a robust method, like the Fibonacci search, with those of a rapid method, like fitting a parabola. Rather than implement a Fibonacci search, Brent uses a simpler to program but slightly less efficient golden section search. For large values of k the ratios of sucessive Fibonacci numbers converge to 0.61803, to five decimals, whose reciprocal is known as the golden ratio. The golden section search uses the constant normalized intervals [0, 0.61803] and [0.38197, 1] to isolate the minimum. Brent's method alternates between a golden section search when the function is refractory, and parabola fitting when the function behaves better. Brent wrote the original algorithm in ALGOL 60, a language little used today. Both Forsythe, Malcolm, and Moler (1977, Sec. 8.2) and Press, Flannery, Teukolsky, and Vetterling (1986, Sec. 10.2) give FORTRAN code for Brent's algorithm.

A line search to determine α, speeds up considerably the steepest descent method. With the function (7.14), a start from $\mathbf{X}^T = (0 \ 0)$, a tolerance of six decimal digits of precision, α determined by a Fibonacci line search, and searching in a direction not of the gradient, but the normalized gradient

$g(\mathbf{X})/\|g(\mathbf{X})\|_2$, sometimes called the direction number, the steepest descent method converges with thirty-one iterations, instead of nearly two thousand. But the DFP method does even better, just three iterations. The first iteration coincides with the steepest descent iteration, but $\mathbf{H}_i^{-1}$ converges so quickly to the inverse Hessian that convergence accelerates markedly.

It would be possible to include a relatively short FORTRAN program to illustrate the DFP method, Eqns. (7.23)–(7.28), but such a program would hardly be efficient. The matrices $\mathbf{H}_i^{-1}$, $\mathbf{A}_i$, and $\mathbf{B}_i$ are all symmetric, and advantage should be taken of this. Instead of using three arrays of size $n \times n$, we should use arrays of size $n(n + 1)/2$. In fact, practical algorithms for the DFP method, such as IBM's SSP package, do even better and use one array of size $n(n + 7)/2$. But the resulting program, although more efficient, is also longer and more difficult to understand and will not be given here.

Given the advantages of the DFP method over that of steepest descent, we would hardly choose the latter. But both are methods for nonlinear optimization, unconstrained optimization, applicable to any nonpathological function. If our interest is nonlinear least squares, generally referred to as the nonlinear regression problem, Eq. (7.9) assumes the special form

$$F(\mathbf{X}) = \mathbf{p}^T(\mathbf{X}) \cdot \mathbf{p}(\mathbf{X}), \tag{7.33}$$

with

$$\mathbf{p}(\mathbf{X}) = \begin{pmatrix} h_1(\mathbf{X}) - d_1 \\ h_2(\mathbf{X}) - d_2 \\ \vdots \\ h_n(\mathbf{X}) - d_n \end{pmatrix},$$

and if advantage is taken of this special form a more efficient method may result.

The gradient of Eq. (7.33) is

$$\mathbf{g}(\mathbf{X}) = 2\mathbf{J}^T(\mathbf{X}) \cdot \mathbf{p}(\mathbf{X}), \tag{7.34}$$

where $\mathbf{J}(\mathbf{X})$ is the Jacobian matrix of partial derivatives

$$\mathbf{J}(\mathbf{X}) = \begin{pmatrix} \dfrac{\partial p_1}{\partial X_1} & \dfrac{\partial p_1}{\partial X_2} & \cdots & \dfrac{\partial p_1}{\partial X_n} \\[2mm] \dfrac{\partial p_2}{\partial X_1} & \dfrac{\partial p_2}{\partial X_2} & \cdots & \dfrac{\partial p_2}{\partial X_n} \\[2mm] \vdots & & & \\[2mm] \dfrac{\partial p_m}{\partial X_1} & \dfrac{\partial p_m}{\partial X_2} & \cdots & \dfrac{\partial p_m}{\partial X_n} \end{pmatrix}. \tag{7.35}$$

Suppose that we are at point $\mathbf{X}_i$ and wish to move to point $\mathbf{X}_{i+1}$ such that Eq. (7.33) decreases. Develop each component $p_i(\mathbf{X})$ of $\mathbf{p}(\mathbf{X})$ in a Taylor series about X_i and truncate the series after the first derivative term. There are m such equations which may be expressed in matrix notation as

$$\mathbf{p}(X_{i+1}) = \mathbf{p}(X_i) + \mathbf{J}(X_i) \cdot (X_{i+1} - X_i). \tag{7.36}$$

Notice the difference between Eqns (7.36) and (7.15). Unlike the latter the gradients in $\mathbf{J}(X_i)$ do not vanish. The gradient we want to vanish is Eq. (7.34), not the individual gradients in the Jacobian.

We hope that X_{i+1} is a better approximation to the solution than X_i and that the gradient $\mathbf{g}(X_{i+1})$ vanishes,

$$\mathbf{g}(X_{i+1}) = \mathbf{J}^T(X_{i+1}) \cdot \mathbf{p}(X_{i+1}) = 0. \tag{7.37}$$

Substitute Eq. (7.36) into Eq. (7.37) and, because the two solutions X_{i+1} and X_i are assumed close to one another (if that were not true we would be unjustified in truncating the Taylor series after the linear term), set $\mathbf{J}^T(X_{i+1}) \cong \mathbf{J}^T(X_i)$. We obtain

$$X_{i+1} = X_i - [\mathbf{J}^T(X_i) \cdot \mathbf{J}(X_i)]^{-1} \cdot \mathbf{J}^T(X_i) \cdot \mathbf{p}(X_i), \tag{7.38}$$

known as the Gauss–Newton method for solving a nonlinear equation. The nonlinear equation is replaced by a series of linear equations.

Because the Jacobian is evaluated at a specific point X_i, as is $\mathbf{p}(X_i)$, Eq. (7.38) is nothing more than an equation embodying linear least squares: the term $(\mathbf{J}^T \cdot \mathbf{J})^{-1} \cdot \mathbf{J}^T$ is our old friend the pseudoinverse from Section 5.1 and $\mathbf{J}^T \cdot \mathbf{J}$ are the normal equations. Any of the methods discussed in Chapter 5 may be used to solve Eq. (7.38). It is even unnecessary to form normal equations. Let $\Delta X_i = X_{i+1} - X_i$ and write Eq. (7.38) as

$$\mathbf{J}^T \cdot \mathbf{J} \Delta X_i = -\mathbf{J}^T \cdot \mathbf{p}. \tag{7.39}$$

Multiply both sides by $\mathbf{J}^{-T}$ (interpreted as a pseudoinverse) to obtain

$$\mathbf{J}(X_i) \cdot \Delta X_i = -\mathbf{p}(X_i), \tag{7.40}$$

the same form as Eq. (4.1). We could, if we so wished, employ orthogonal transformations to calculate the solution ΔX_i.

In general, the Gauss–Newton method must be iterated more than once before convergence to a specified accuracy is achieved, if indeed the iterates even converge. At the solution the Jacobian matrix may be calculated, from which the normal equations and hence the variance–covariance and correlation matrices and error estimates for the solution all follow from the methods of Chapter 5.

To illustrate the Gauss–Newton method consider the solution of Eq. (7.6). The function to be minimized is

$$F(X_i, X_2) = (X_1 - 1)^2 + (X_1 e^{X_2} - 2)^2 + (X_1 e^{2X_2} - 6)^2, \tag{7.41}$$

with

$$\mathbf{p}^T = (X_1 - 1 \quad X_1 e^{X_2} - 2 \quad X_1 e^{2X_1} - 6)$$

and

$$\mathbf{J} = \begin{pmatrix} 1 & 0 \\ e^{X_2} & X_1 e^{X_2} \\ e^{2X_2} & 2X_1 e^{2X_2} \end{pmatrix}.$$

Take as the first approximation the solution $\mathbf{X}^T = (0.93466 \quad 0.89588)$ given by the logarithmic transformation. We calculate for Eq. (7.39), working to five decimals,

$$\begin{pmatrix} 43.00004 & 72.90355 \\ 72.90355 & 131.03853 \end{pmatrix} \begin{pmatrix} \Delta X_1 \\ \Delta X_2 \end{pmatrix} = \begin{pmatrix} 1.70858 \\ 3.73440 \end{pmatrix},$$

whose solution $\Delta \mathbf{X}^T = (-0.15125 \quad 0.11265)$ results in a second approximation of $\mathbf{X}^T = (0.78341 \quad 1.00853)$. Five iterations suffice to obtain a precision of five decimals in the solution. At the solution of $\mathbf{X}^T = (0.78044 \quad 1.01742)$ the normal equations are

$$\begin{pmatrix} 67.18925 & 97.34232 \\ 97.34232 & 147.27954 \end{pmatrix},$$

which give a variance–covariance matrix of

$$\begin{pmatrix} 0.35060 & -0.23172 \\ -0.23172 & 0.15995 \end{pmatrix}$$

and a correlation matrix of

$$\begin{pmatrix} 1.00000 & -0.97851 \\ -0.97851 & 1.00000 \end{pmatrix}.$$

From the sum of the squares of the residuals, 0.07423, we calculate $\sigma(1) = 0.27246$ and the errors of the solution as $\mathbf{\varepsilon}^T = (0.16133 \quad 0.10897)$. Notice that, unlike the situation with the linear least squares of Chapter 5, the elements of the matrix of the normal equations, and therefore the correlations and errors of the unknowns, vary as the solution varies.

Under what conditions does the Gauss–Newton method converge? Forsythe, Malcolm, and Moler (1977, p. 170) show that if $\mathbf{X}_i$ is sufficiently close to $\mathbf{X}_{i+1}$ the iterates converge and converge quadratically. But how close is close? This depends on the structure of the term $(\mathbf{J}^T \cdot \mathbf{J})^{-1} \cdot \mathbf{J}^T \cdot \mathbf{p}$. The question of convergence is complicated even more by noise in the data coupled with the granularity of the floating-point number system. We may be close to the minimum, but the derivatives in the Jacobian matrix may be so inaccurate because of experimental noise that Eq. (7.39) moves the solution away from the minimum.

We find, in practice, that the Gauss–Newton method does one of three things: it quickly converges to a minimum; it quickly diverges to infinity; or it does neither, but calculates solutions that wander about. Quick divergence to infinity—which on a computer means floating-point overflow—is usually brought about by a poor initial approximation to the solution and wandering by noisy data.

The Gauss–Newton method is best reserved for the times when we know that our initial approximation is good. Trajectory calculations, for example, use a preliminary trajectory that is corrected as further information becomes available. A space probe nearing a planet has its trajectory corrected from the probe–planet distance as soon as this information can be determined. Astronomers refer to the process as the differential correction of an orbit. The name itself implies that the preliminary orbit is good and our main interest is tweaking it a bit to improve accuracy.

Part of the difficulty with the convergence of the Gauss–Newton method arises from the Jacobian matrix being ill-conditioned. Both Levenberg and Marquardt proposed a way to improve the conditioning of the problem, thereby improving convergence. Start with Eq. (7.21), the basic iterative equation for the DFP method, but instead of approximating the inverse Hessian matrix, let us calculate the Hessian matrix directly. From Eq. (7.33) we obtain

$$\frac{\partial F}{\partial X_j} = 2 \sum_{i=1}^{m} (h_i(\mathbf{X}) - d_i) \frac{\partial h_i}{\partial X_j} \tag{7.42}$$

and

$$\frac{\partial^2 F}{\partial X_j \partial X_k} = 2 \sum_{i=1}^{m} (h_i(\mathbf{X}) - d_i) \frac{\partial^2 h_i}{\partial X_j \partial X_k} + 2 \sum_{i=1}^{m} \frac{\partial h_i}{\partial X_j} \frac{\partial h_i}{\partial X_k}. \tag{7.43}$$

Denote the Hessian matrix of Eq. (7.33), whose elements are Eq. (7.43), by $\mathbf{H}(\mathbf{X})$ and the Hessian matrix of $p_i(\mathbf{X}) = h_i(\mathbf{X}) - d_i$ by $\mathbf{h}(\mathbf{X})$. Then Eq. (7.43) becomes

$$\mathbf{H}(\mathbf{X}) = 2\mathbf{J}^{\mathrm{T}}(\mathbf{X}) \cdot \mathbf{J}(\mathbf{X}) + 2\mathbf{p}(\mathbf{X}) \cdot \mathbf{h}(\mathbf{X}). \tag{7.44}$$

To parallel the derivation of Eq. (7.21) substitute Eqns. (7.34) and (7.44) into Eq. (7.17). We obtain

$$\mathbf{X}_{i+1} = \mathbf{X}_i - \mathbf{H}^{-1}(\mathbf{X}_i) \cdot \mathbf{g}(\mathbf{X}_i). \tag{7.45}$$

If the second term on the right-hand side of Eq. (7.44) were neglected, Eq. (7.45) would coincide with Eq. (7.38). But why should such a term, derived from exact mathematics, be omitted? Because $\mathbf{p}(\mathbf{X})$ is the m vector of the residuals. If discordant data are present, and in realistic problems they invariably are, inclusion of the term prevents Eq. (7.44) approximating well the Hessian matrix of Eq. (7.33), and may cause the algorithm to diverge. If discordant data have been filtered out, then the term should be included.

Whether the second term on the right-hand side of Eq. (7.44) is included or omitted, Marquardt (1963), following an original suggestion of Levenberg (1944), modifies Eq. (7.45) to

$$\mathbf{X}_{i+1} = \mathbf{X}_i - (\mathbf{H}(\mathbf{X}_i) + \lambda \mathbf{I})^{-1} \cdot g(\mathbf{X}_i), \tag{7.46}$$

where $\mathbf{I}$ is the unit matrix and λ is a nonnegative constant that improves the conditioning of the problem. (The factor α of Eq. (7.21) is unnecessary in Eq. (7.46) because λ handles the scale problem.) When λ is small, Eq. (7.46) approximates the Gauss–Newton method and when λ is large the steepest descent method. In general, we want λ as small as possible to accelerate convergence, but its determination presents problems.

Press, Flannery, Teukolsky, and Vetterling (1986, pp. 523–528) proceed by picking a small but arbitrary value for λ, such as 0.001, and solving the linear system, Eq. (7.46), for the next $\mathbf{X}_{i+1}$. If the sum square of the residuals with the new solution increases, the proposed λ is rejected, a larger λ selected, and the solution recalculated. If, however, the sum square of the residuals decreases with the solution $\mathbf{X}_{i+1}$, the solution is accepted, but λ nevertheless decreased for the next iteration. We keep repeating the process until convergence is achieved. Notice that selection of λ involves the solution of at least one, and possibly many, linear systems with each iteration.

Alternatively (Kennedy and Gentle, 1980, pp. 444–445), we may perform an eigenvalue–eigenvector decomposition of the matrix $\mathbf{H}(\mathbf{X}_i)$ (see Eq. (2.29)). An efficient decomposition, however, does not result from use of the definition, but from the QR algorithm presented in the next chapter. The decomposition gives

$$\mathbf{H} = \mathbf{V}^{\mathrm{T}} \cdot \mathbf{S}^2 \cdot \mathbf{V}, \tag{7.47}$$

where $\mathbf{S}^2$ is a diagonal matrix of the eigenvalues—it is written as $\mathbf{S}^2$ rather than $\mathbf{S}$ to conform with the notation in the next chapter—and $\mathbf{V}$ is an orthogonal matrix of the eigenvectors. We select for λ the smallest value such that $S_j^2 + \lambda > \varepsilon$, where ε is the machine epsilon of the precision used, for all values of $j = 1, 2, \ldots, n$. Then the inverse matrix in Eq. (7.46) is expressed by the backwards product expansion

$$(\mathbf{H} + \lambda \mathbf{I})^{-1} = \sum_{j=1}^{n} (S_j^2 + \lambda)^{-1} \mathbf{v}_j \cdot \mathbf{v}_j^{\mathrm{T}}, \tag{7.48}$$

where $\mathbf{v}_j$ is the jth column of $\mathbf{V}$.

Performing an eigenvalue–eigenvector decomposition with every iteration is expensive. Even the efficient QR algorithm needs about $\frac{2}{3}n^3 + 44n^2$ operations to find all of the eigenvalues and eigenvectors. Hunting for the right λ by solving several linear systems, at $n^3/6$ operations per Cholesky decomposition, would probably be more efficient. But it may also fail. For poorly conditioned problems, or a poor initial starting solution, the Hessian matrix may not be positive definite or even exist at all, and Cholesky decomposition, the most efficient way to solve Eq. (7.46), comes to grief. An eigenvalue–eigenvector decomposition calculates a λ guaranteed to avoid a poorly conditioned matrix. Kennedy and Gentle (1980, pp. 445–450) present additional methods for poorly conditioned matrices.

Code for the Levenberg–Marquardt method is complex, particularly if we use an eigenvalue–eigenvector decomposition to compute λ, and will not given here. Press, Flannery, Teukolsky, and Vetterling (1986, pp. 526–528) present relatively short FORTRAN subroutines for the Levenberg–Marquardt method, as implemented by finding λ from solving a series of linear systems. But really efficient code, that takes advantage of the symmetry of the Hessian matrix and does extensive error checking, needs several hundred lines, not including comments.

Both the DFP and the Levenberg–Marquardt are efficient nonlinear methods. For work in the L_2 norm, the nonlinear regression problem, the latter would normally be chosen as more efficient: we calculate the inverse Hessian directly rather than starting out with a unit matrix that more closely approximates the inverse Hessian with successive iterations. But the DFP method can be employed with other norms. Abdelmalek (1971) has used the method to find linear L_1 solutions. Although the Barrodale and Roberts algorithm of Chapter 6 is far more efficient for this purpose, the DFP method can just as easily be adopted for nonlinear L_1 solutions. Hald and Madsen (1985) use a combined linear programming and quasi-Newton—similar to but not identical with the DFP method—for nonlinear L_1 optimization. For strict L_1 solutions gradient methods must be applied with care because as the p of Eq. (7.9) approaches unity the components of the gradient also approach unity in absolute value—recall Eqns. (6.15)–(6.18). The usual trick calculates a solution with a power such as $p = 1.5$, uses the solution with a lower power such as $p = 1.2$, and so forth, down to p almost but not quite equal to unity, say $p = 1.01$.

Even with the nonlinear regression problem, if we suspect that discordant data are present it may be worthwhile to calculate the first solution by the DFP method using some power of p such as $p = 1.5$ for greater robustness. After we filter out the discordant data indicated from a calculation of the residuals from the DFP solution, we can switch to the least squares criterion and the Levenberg–Marquardt method. In Chapter 6 we emphasized calculating a solution to a linear problem by the L_1 algorithm, followed, if necessary, by a least squares solution or, alternatively, achieving robustness by use of iteratively reweighted least squares. With linear systems a solution can always be calculated. But a nonlinear least squares method may not converge at all in the presence of badly discordant data. Therefore, an iteratively reweighted nonlinear least squares approach may not be possible, and a first robust solution should be obtained by some other method such as DFP with a low power of p.

7.3. Nongradient Methods

Gradient methods suffer from having to calculate the gradient, and sometimes the Hessian matrix, which for refractory functions can be extremely time-consuming. We may always, of course, use a finite difference approximation to the gradient. From the mathematical definition of the partial derivative we may approximate a component of the gradient from

$$\frac{\partial f(X_1, X_2, \ldots, X_n)}{\partial X_j} \cong \frac{f(X_1, X_2, \ldots, X_j + \Delta X_j, \ldots, X_n) - f(X_1, X_2, \ldots, X_n)}{\Delta X_j}$$

$$(7.49)$$

with an error in the approximation proportional to ΔX_j^2. A more accurate, but computationally more demanding, approximation is given by

$$\frac{\partial f(X_1, X_2, \ldots, X_n)}{\partial X_j} \tag{7.50}$$

$$\cong \frac{f(X_1, X_2, \ldots, X_j + \Delta X_j, \ldots, X_n) - f(X_1, X_2, \ldots, X_j - \Delta X_j, \ldots, X_n)}{2\Delta X_j},$$

whose error term is proportional to ΔX_j^3. Selection of ΔX_j must be done carefully; if it is too large the approximation to the partial derivative is poor, and if it is too small precision is lost in the subtraction of two nearly equal quantities.

Even when gradients, whether analytic or numeric, are available, gradient methods are prone to divergence. Noise in the data exacerbates the problem because differentiation is a noise amplifier. And convergence may be to a nearby local minimum rather than the desired global minimum.

As an illustration of the latter, the author once worked on a nonlinear problem in astronomy. The equation to be minimized was

$$\sum_{i=1}^{m'} \cos^2 \delta_c (\alpha_o - \alpha_c)^2 + \sum_{i=1}^{m''} (\delta_o - \delta_c)^2 = \min, \tag{7.51}$$

where α and δ are angles—the subscript o refers to an observed angle—m' is the number of equations involving α, and m'' is the number involving δ. It is unnecessary to write down the complete expressions for α_c and δ_c, which are inverse tangents whose arguments incorporate six and nine variables, respectively. The important observation is that Eq. (7.51) is clearly bimodal. Unless the initial starting solution is close to the true solution a gradient method could easily converge, if it converges, to the nearby minimum, which may not be the global minimum of Eq. (7.51).

Nongradient methods, of which many have been proposed, are far less prone to divergence—some of them, in fact, always converge—and can be made to converge (when they converge), to a global minimum. But they are also almost invariably slower than gradient methods because they make no use of the information supplied by the gradient and, for some methods, the Hessian matrix.

Let us refer to Eq. (7.9), the sum of the pth power of the absolute values of the residuals, by the convenient terminology of "response function". The simplest method conceptually establishes a mesh of points in the n-dimensional space, so many points that the global minimum is bound to be somewhere in the mesh, and calculates the value of the response function at each point. Then a mesh with finer resolution is created near the minimum and the process repeated. This is done as many times as needed to decrease the resolution of the mesh to less than a pre-established tolerance.

Consider, for example, the minimization of Eq. (7.41). Start by calculating the response function for X_1 running from -10 to 10 in steps of one and the same for X_2. The following table shows the progress of the scheme (R is the resolution of the mesh, F the value of the response function) for a specified tolerance of 0.0000001:

X_1	X_2	R	F
1.00000	1.00000	1.00000	2.44541
0.80000	1.00000	0.10000	0.07837
0.79000	1.01000	0.01000	0.07467
0.78100	1.01700	0.00100	0.07424
0.78050	1.01740	0.00010	0.07424
0.78044	1.01743	0.00001	0.07424
0.78044	1.01743	0.000001	0.07424

The final results are the same, to within round-off error, as those found previously from the Gauss–Newton method, but it is unnecessary to specify an initial approximation to start the process.

The mesh method is robust, easy to program (forty-five lines of FORTRAN for the example just given), converges to a global minimum, and is completely impractical for realistic problems. The example needed 3,104 function evaluations to meet the tolerance. The basic difficulty is that the method only evaluates the function and pays no attention to the direction in which the function appears to be moving, information supplied by the gradient and the Hessian in gradient methods. Such lack of finesse becomes intolerable as the number of dimensions increases. Even dedicated number crunchers will be appalled at the function evaluations needed within DO loops nested n deep.

The method can, however, be used to find a good starting region near the global minimum for a gradient method. Provided the resolution is not too fine the mesh method will isolate such a region with a reasonable amount of computing. The gradient method then need not worry about the distraction of local minima, although it can still be derailed by noisy data.

An uncomplicated modification maintains the robustness of the mesh method and greatly increases its efficiency without turning it into a gradient method. Nelder and Mead (1965) proposed a simplex method for nonlinear optimization, which can also be used for nonlinear regression. The method, having no connection with Dantzig's simplex algorithm for linear programming, employs a construct from topology called a simplex. Early versions of Dantzig's algorithm used the same construct, from which the name "simplex algorithm" was derived.

Any textbook on topology—the author's favorite is Patterson (1969) because it is brief, inexpensive, and unencumbered by excessively abstract mathematical detail—discusses a simplex. Let P_0, P_1, ..., P_n be $n + 1$ points in a Euclidean space of n dimensions. An n simplex is the set of points X

defined by

$$\mathbf{X} = \theta_0 \mathbf{P}_0 + \theta_1 \mathbf{P}_1 + \cdots + \theta_n \mathbf{P}_n, \tag{7.52}$$

where

$$\sum_{i=0}^{n} \theta_i, \qquad \theta_i \geq 0.$$

Thus, in a space of zero dimensions a simplex is a point, in a space of one dimension a straight line, in two dimensions a triangle, in three dimensions a tetrahedron, and so on for spaces of higher dimension. If some of the points are collinear or coincide the simplex is degenerate. Only nondegenerate simplexes are of use in the method about to be described.

Because we are fitting n variables our simplex has $n + 1$ vertices. The simplex method starts out by initializing the vertices. One of the vertices is the initial approximation to the solution. Unlike gradient methods the simplex method is insensitive to the initial approximation. Nearly any value will do and $\mathbf{X}^T = (0 \ 0 \ \ldots \ 0)$ is often chosen. The other vertices are set to the initial approximation varied by an amount λ reasonable for the problem at hand,

$$\mathbf{P}_i = \mathbf{P}_0 + \lambda \mathbf{e}_i, \qquad i = 1, \ldots, n, \tag{7.53}$$

where $\mathbf{P}_0$ is the initial approximation, $\mathbf{P}_i$ the other vertices, and $\mathbf{e}_i$ the unit vectors of Eq. (2.44). Notice that if λ is unity the $\mathbf{P}_i$'s of Eq. (7.53) concide with the $\mathbf{P}_i$'s of Chapter 6 for an initial approximation of $\mathbf{P}_0$ equal to the null vector.

The simplex method calculates the response function at each vertex, rejects the worst—the one with the highest value—and calculates a new vertex to replace it in such a way that the nondegeneracy of the simplex is maintained. Four mechanisms for the calculation of the new vertex are used: reflection, expansion, contraction along one dimension, and contraction along all dimensions (shrinkage). Figure 7.3 illustrates these four ways using the equilateral triangle given by Eq. (7.53) for the first iteration of the simplex method when two variables are involved.

Suppose that the origin happens to have the highest response H; L is the lowest response and I an intermediate response. The initial simplex is L–H–I. We want to reject H. The simplex method first calculates the barycenter, or center of mass, of the other vertices, point B, and reflects H about the line H–B by the same distance as H–B to create R. Notice that the areas of the two triangles L–H–I and L–I–R are the same; degeneracy is not allowed.

The method tests the response at R. Several outcomes are possible. Because we are interested in minimizing the response function, having R lower than any of the previous responses L, H, I is the best outcome. We are obviously moving in the correct direction. The method, therefore, goes even farther and calculates an expanded vertex E along the line H–B–R. The actual distance R–E is somewhat arbitrary; we typically take the distance R–E the same as B–R, or twice the distance H–B. If E is better than R it is accepted, the new simplex becomes L–I–E, and the current iteration terminates; otherwise, if R is better than E the simplex is L–I–R.

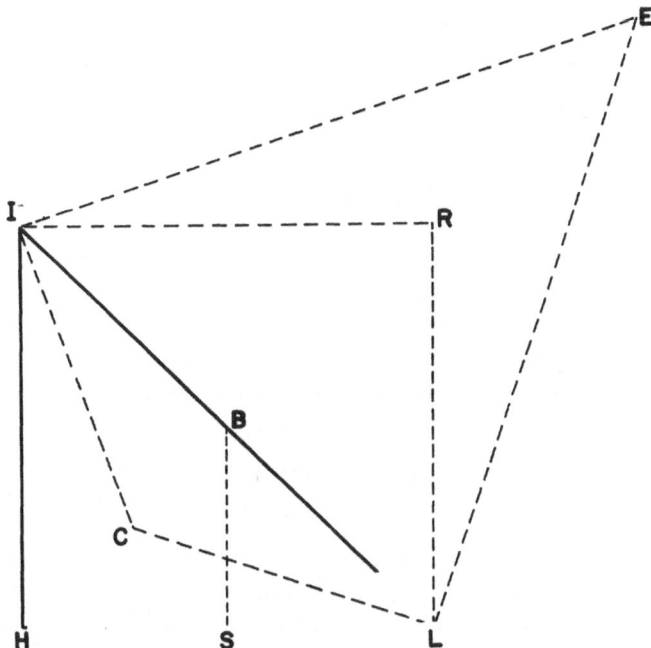

Figure 7.3. Equilateral triangle L–H–I for initial simplex: H = vertex with highest response, L = vertex with lowest response, I = intermediate vertex, B = barycenter of all but worst vertices, R = reflected vertex, E = expanded vertex, C = contracted vertex, S and B = shrunken vertices.

If the response at R is better, or at least no worse, than the response at H, but worse than that at L, then R is accepted without trying an expansion. The simplex is updated to L–I–R, and the current iteration terminates.

But if the response at R is worse than the response at H, the selection is not accepted. Instead, the method tries a contraction along the line H–B and calculates the response at C, typically half way between H and B. If C gives a lower value for the response function than H, then the new simplex becomes L–C–I and the current iteration terminates. But if C is worse, or at least no better, than H, the entire simplex is contracted along all of the directions towards the point L with the lowest value of the response function. Typically each distance, here L–H and L–I, is shrunk by half. The new simplex becomes L–S–B, and the current iteration terminates. Notice that each iteration of the simplex method requires between $n + 1$ and $2(n + 1)$ function evaluations.

To illustrate these ideas in a concrete setting let us return to Eq. (7.41). We use a matrix of size $(n + 1) \times (n + 1)$ to hold the simplex. Each row of the matrix contains the coordinates of a vertex in the first n columns and the response function in the $(n + 1)$st column. Take $\lambda = 1$ and use $\mathbf{X}^T = (0\ 0)$ as the first approximation. The starting simplex is

$$\begin{pmatrix} 0 & 0 & 41.00000 \\ 0 & 1 & 41.00000 \\ 1 & 0 & 26.00000 \end{pmatrix}.$$

Both the origin, H of Figure 7.3, and the intermediate point I have the same value for the response function. We will, somewhat arbitrarily, take the origin as the vertex to be replaced.

Reflecting about the barycenter, coordinates (0.5 0.5), we find R at co-ordinates (1 1) with response function 2.44541, a substantial improvement. Therefore, we attempt an expansion. E has coordinates (1.5 1.5) and response function 604.72744; E is rejected, R accepted, and the simplex updated to

$$\begin{pmatrix} 1 & 1 & 2.44541 \\ 0 & 1 & 41.00000 \\ 1 & 0 & 26.00000 \end{pmatrix}:$$

The new simplex is L–I–R of Figure 7.3. Point I has the worst response and will be eliminated. Reflecting about the point midway between R and L we find the new reflected point to the right with coordinates (2 0) and response function 17.00000. This is better than I but worse than R. No expansion is tried and the simplex updated to

$$\begin{pmatrix} 1 & 1 & 2.44541 \\ 2 & 0 & 17.00000 \\ 1 & 0 & 26.00000 \end{pmatrix}.$$

We will follow the example for one more iteration. Call the point (2 0) R'. Indicated is a rejection of point I. Reflecting about the line R–R' we calculate a new reflected point R'' with coordinates (2 1) and response function 89.86522. Bad. R'' is rejected and a contraction along the line L–R'' tried. Figure 7.4 shows the situation. C has coordinates (0.25 1.25) and response function 15.73500. The point C is selected to replace L. The new simplex, R'–C–R, is

$$\begin{pmatrix} 1 & 1 & 2.44541 \\ 2 & 0 & 17.00000 \\ 0.25 & 1.25 & 15.73500 \end{pmatrix}.$$

After sixty-six iterations convergence to a tolerance of 0.0000001 is reached, the method performing four expansions, fifty-one contractions, and no shrink-ages. There is, therefore, a total of 363 function evaluations: three each for sixty-six reflections, fifty-one contractions, and four expansions. This compares well with the 3,104 evaluations needed by the mesh method with the same tolerance. As with gradient methods the tolerance should be no smaller than the square root of the machine epsilon for the precision used when we monitor the decrease in the response function. When monitoring the variables them-

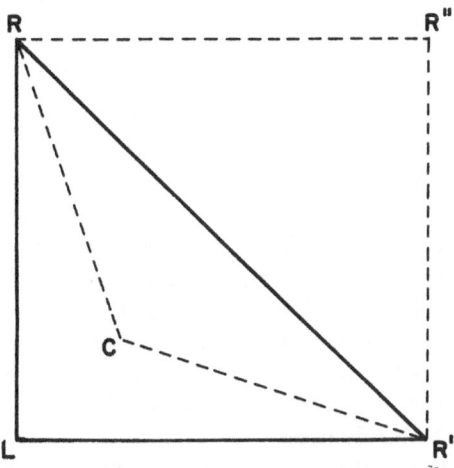

Figure 7.4. Third iteration of simplex method for Eq. (7.41) and initial approximation $\mathbf{X}^T = (0 \quad 0)$.

selves the tolerance may be set to something of the order of the machine epsilon.

How do we know that the simplex method will, in a given application, converge to a global minimum? We don't. But once we have a minimum we can restart the simplex method there, preferably with a larger λ. Convergence to the same minimum means the minimum is most likely global; otherwise, the simplex would wander off in a new direction and converge to a different minimum. Alternatively, we could use the mesh method to isolate the global minimum and start the simplex method in this region.

Figure 7.5 presents a FORTRAN program for the simplex method, based on Caceci and Cacheris's (1984) PASCAL program, but with some significant differences. The most notable difference, of course, is the use of FORTRAN rather than PASCAL, a language hardly suited to serious numerical computation. But translation from one language to another, particularly structured languages like FORTRAN-77 and PASCAL, is straightforward and can even be automated.

More significantly, the program in Figure 7.5 can be used to minimize any power of the sum of the residuals. If we were to set $p = 1$, we would obtain a nonlinear L_1 solution. When we opt for a least squares solution, $p = 2$, the program calculates the mean errors of the solution and the correlation matrix by forming a variance–covariance matrix at the solution point. The statement CALL NORSLVE(A, N) transfers control to a subroutine, basically the program of Figure 5.1 in subroutine form with the lines for solving the normal equations omitted, to invert the matrix, use the diagonal elements for the calculation of the errors, and compute the correlations. Because the subroutine parallels Figure 5.1 so closely the code is not repeated. Caceci and

Cacheris (1984) find the errors by Monte Carlo simulation, computationally expensive, as Section 6.4 points out, and hardly justifiable if we can easily calculate a variance–covariance matrix. Nonleast squares norms, $p \neq 2$, have no associated variance–covariance matrix, and the program calculates no errors or correlations. For these norms we would have to use a Monte Carlo simulation for the errors, although for an L_1 solution Eq. (6.58), with A and A_j^T found from Jacobians evaluated at the solution, could be used.

Figure 7.5 tests for the situation $m < n$, an underdetermined system, and in such an instance terminates immediately. The simplex is initialized by Eq. (7.53), although other initializations, such as the one of Caceci and Cacheris (1984), are possible. The program's function subroutine minimizes Eq. (7.41). For other functions a few lines of the subroutine and the lines in the main program that calculate the partial derivatives for the variance–covariance matrix would have to be changed.

The simplex method, although more efficient than the atrociously slow mesh method, still lags behind gradient methods in terms of efficiency. The only exception may be when high correlations among the unknowns occur. In some examples with high correlations the author has been the simplex method outperform the DFP method, but not the Levenberg–Marquardt method.

But, in general, the simplex method is too time-consuming to be recommended when something faster is available. As a concrete example the author once applied the simplex method to a nonlinear problem involving 12,073 equations of condition and five unknowns. Over thirty hours of CPU time were needed to achieve three decimal precision in the response function. The Gauss–Newton method tackled the same problem in fifty minutes.

And something faster is always available—a gradient method with, if necessary, numerically calculated partial derivatives. Unfortunately, it may not work. Even when we use the mesh method to isolate the global maximum, avoiding the difficulty of convergence to a local minimum, divergence may occur for sufficiently noisy data. The author has seen this happen when processing astronomical images. Even when the initial approximation was close to what was eventually accepted as the solution, the Gauss–Newton iterates sometimes diverged rapidly to infinity and sometimes merely wandered when the image was noisy. The DFP method was also prone to divergence, but the simplex method always converged.

These are the occasions, assured convergence, when the simplex method becomes the tool of choice, even if we pay a high computational price. But we should still try to estimate as many of the variables as possible by faster methods, leaving only the most refractory ones for the slow but sure simplex method.

Other techniques for gradient-free nonlinear optimization, applicable also to nonlinear regression, have been developed. Two of the more popular ones are Rosenbrock's (1960) method and Powell's, of DFP fame, conjugate direction method (Acton, 1970, pp. 464–467). According to some sources the conjugate direction method may be faster than the simplex method, but the author has not tried it and makes no claims.

This chapter has concerned itself exclusively with unconstrained optimization and regression, the usual situation in practice. Those interested in nonlinear optimization with equality constraints will find Intriligator (1981, pp. 59–65) useful; the same source (pp. 66–72) also discusses optimization with inequality constraints.

```
C
C
C     SUBROUTINE FOR DOING A FIBONACCI SEARCH FOR A FUNCTION MINIMUM IN THE
C     DIRECTION DEFINED BY THE DIRECTION NUMBER AT POINT X .
C
C
            SUBROUTINE  FIB-SRCH ( X , D , ALPHA , N )
            IMPLICIT REAL * 8  ( A - H , O - Z )
            REAL * 8  LARGE
            DIMENSION  X ( N ) , D ( N ) , XHOLD ( 20 )
            COMMON  / A /  TOLER
            DO  I = 1 , N
              XHOLD ( I ) = X ( I )
            END  DO
            ALPHA = 0.5D0
            FOLD = FUNC ( X , N )
            F = FOLD - 1D0 / TOLER
C
C     BRACKET THE FUNCTION MINIMUM .
C
            DO  WHILE  ( F . LT . FOLD )
              FOLD = F
              DO  I = 1 , N
                X ( I ) = XHOLD ( I ) - ALPHA * D ( I )
              END  DO
              F = FUNC ( X , N )
              ALPHA = ALPHA * 2D0
            END  DO
C
C     CALCULATE FIBONACCI NUMBER GREATER THAN TOLERANCE .
C
            DO  I = 1 , 1000
              IF  ( FIB ( I ) . GT . TOLER ) GOTO 5
            END  DO
5           K = I
C
C     USE FIBONACCI SEQUENCE BETWEEN LIMITS A  AND  B  TO FIND NEW FUNCTION MINIMUM .
C     SMALL IS THE SMALLER OF THE INTERVALS , LARGE THE LARGER . THE FIBONACCI IN-
C     TERVAL  0 , 1   IS NORMALIZED TO   A , B    . ALPHA IS THE DISTANCE TO THE MINIMUM .
C
            A = 0D0
            B = ALPHA
            LIM = K
            DO  I = LIM , 2 , -1
              DIFF = B - A
```

Figure 7.2. Subroutine for Fibonacci search.

```
                    SMALL = A + ( FIB ( K - 2 ) / FIB ( K ) ) * DIFF
                    LARGE = A + ( FIB ( K - 1 ) / FIB ( K ) ) * DIFF
                    DO J = 1, N
                       X ( J ) = XHOLD ( J ) - SMALL * D ( J )
                    END DO
                    FSMALL = FUNC ( X , N )
                    DO J = 1, N
                       X ( J ) = XHOLD ( J ) - LARGE * D ( J )
                    END DO
                    FLARGE = FUNC ( X , N )
                    IF ( FSMALL . LT . FLARGE ) THEN
                       B = LARGE
                    ELSE
                       A = SMALL
                    END IF
                    K = K - 1
                    ALPHA = ( SMALL + LARGE ) / 2D0
                 END DO
C
C     CALCULATE NEW VALUES FOR THE X'S .
C
                 DO I = 1, N
                    X ( I ) = XHOLD ( I ) - ALPHA * D ( I )
                 END DO
                 RETURN
                 END
C
C
C     FUNCTION TO CALCULATE FIBONACCI NUMBERS .
C
C
                 FUNCTION FIB ( N )
                 REAL * 8 FIB , F0 , F1 , FN
                 IF ( N . LT . 0 ) STOP ' N MUST BE NON - NEGATIVE '
                 F0 = 0.0
                 F1 = 1.0
                 IF ( N . EQ . 0 . OR . N . EQ . 1 ) THEN
                    IF ( N . EQ . 0 ) FN = F0
                    IF ( N . EQ . 1 ) FN = F1
                    FIB = FN
                    RETURN
                 ELSE
                    DO I = 2, N
                       FN = F0 + F1
                       F0 = F1
                       F1 = FN
                    END DO
                    FIB = FN
                 END IF
                 RETURN
                 END
```

Figure 7.2 (*continued*)

```
C
C
C    PERFORM A NONLINEAR ADJUSTMENT OF ANY POWER OF THE RESIDUALS BY USING NELDER
C    AND MEAD'S SIMPLEX ALGORITHM .
C
C
          PROGRAM  SIMPLEX
          IMPLICIT REAL * 8  ( A - H , O - Z )
          REAL * 8  NEXT ( 100 ) , MEAN ( 100 ) , DERV ( 100 ) , A ( 5050 ) , X ( 100 ) , LAMBDA
          INTEGER  H ( 100 ) , L ( 100 )
          LOGICAL  DONE
          DIMENSION  SIMP ( 100 , 100 ) , ERROR ( 100 ) , CENTER ( 100 ) , TEMP ( 100 )
          EQUIVALENCE  ( ERROR , MEAN )
          EQUIVALENCE  ( ERROR , TEMP )
          EQUIVALENCE  ( X , MEAN )
C
C    ALPHA IS THE REFLECTION COEFFICIENT , BETA THE CONTRACTION COEFFICIENT , AND
C    GAMMA THE EXPANSION COEFFICIENT .
C
          DATA  ALPHA / 1.0 /, BETA / 0.5 /, GAMMA / 2.0 /, NITER / 0 /, LUNIT / 1 /
          COMMON  / A /  SIMP , H , L , NEXT , NPL1
          COMMON  / B /  M , POW , LUNIT
          COMMON  / C /  SIGMA
          TYPE * ,  ' HOW MANY PARAMETERS DO YOU WANT TO FIT ?  '
          ACCEPT * , N.
          NPL1 = N + 1
          TYPE * ,  ' HOW MANY EQUATIONS OF CONDITION ARE THERE ?  '
          ACCEPT * , M
          IF  ( M . LT . N )  STOP  ' SYSTEM IS UNDERDETERMINED  '
          TYPE * ,  ' WHAT TOLERANCE DO YOU WANT FOR THE FIT ?  '
          ACCEPT * , TOLER
          TYPE * ,  ' WHICH POWER OF THE RESIDUALS DO YOU WANT TO MINIMIZE ?  '
          ACCEPT * , POW
          TYPE * ,  ' WHAT IS THE LIMIT FOR THE NUMBER OF ITERATIONS ?  '
          ACCEPT * , LIM
          TYPE * ,  ' WHAT IS LAMBDA ?  '
          ACCEPT * , LAMBDA
          OPEN  ( LUNIT , FILE = ' INDATA . DAT ' , STATUS = ' OLD ' )
C
C    INITIALIZE SIMPLEX AND OTHER ARRAYS .
C
          DO  J = 1 , NPL1
            SIMP ( 1 , J ) = 0D0
            DO  I = 2 , NPL1
              IF  ( ( I - 1 ) . EQ . J )  THEN
                SIMP ( I , J ) = LAMBDA
              ELSE
                SIMP ( I , J ) = 0D0
              END  IF
            END  DO
            H ( J ) = J
```

Figure 7.5. Program for nonlinear adjustment of pth power of residuals using simplex method.

```
              L ( J ) = J
              NEXT ( J ) = 0D0
              ERROR ( J ) = 0D0
          END DO
          DO I = 1 , NPL1
            DO K = 1 , N
              TEMP ( K ) = SIMP ( I , K )
            END DO
            SIMP ( I , NPL1 ) = SUMPOWRES ( TEMP , NPL1 )
          END DO
          DO I = 1 , NPL1
            L ( I ) = 1
            H ( I ) = 1
          END DO
C
C   RANK VERTICES .
C
          CALL RANK
C
C   COMPUTE CENTROID EXCLUDING THE WORST .
C
10        DO I = 1 , NPL1
            CENTER ( I ) = 0D0
          END DO
          DO I = 1 , NPL1
            IF ( I . NE . H ( NPL1 ) ) THEN
              DO J = 1 , N
                CENTER ( J ) = CENTER ( J ) + SIMP ( I , J )
              END DO
            END IF
          END DO
C
C   DO THE REFLECTING , CONTRACTING , ETC .
C
          DO I = 1 , NPL1
            CENTER ( I ) = CENTER ( I ) / ( N )
            NEXT ( I ) = ( 1D0 + ALPHA ) * CENTER ( I ) - ALPHA * SIMP ( H ( NPL1 ) , I )
          END DO
          NEXT ( NPL1 ) = SUMPOWRES ( NEXT , NPL1 )
C
C   BETTER THAN THE BEST ?
C
          IF ( NEXT ( NPL1 ) . LE . SIMP ( L ( NPL1 ) , NPL1 ) ) THEN
          CALL NEWVERTEX
C
C   TRY AN EXPANSION .
C
          DO I = 1 , N
            NEXT ( I ) = GAMMA * SIMP ( H ( NPL1 ) , I ) + ( 1D0 - GAMMA ) * CENTER ( I )
          END DO
          NEXT ( NPL1 ) = SUMPOWRES ( NEXT , NPL1 )
```

Figure 7.5 (*continued*)

```
C
C     ACCEPT THE EXPANSION .
C
                  IF  (NEXT ( NPL1 ) . LE . SIMP ( L ( NPL1 ) , NPL1 ) )   CALL NEWVERTEX
C
C     IF NOT BETTER THAN THE BEST . . .
C
              ELSE
                  IF  ( NEXT (NPL1 ) . LE . SIMP ( H ( NPL1 ) , NPL1 ) )  THEN
                      CALL  NEWVERTEX
C
C     BETTER THAN WORST .
C
              ELSE
C
C     WORSE THAN WORST - CONTRACT .
C
                  DO  I = 1 , N
                      NEXT ( I ) = BETA * SIMP ( H ( NPL1 ) , I ) + ( 1D0 - BETA ) * CENTER ( I )
                  END  DO
                  NEXT ( NPL1 ) = SUMPOWRES ( NEXT , NPL1 )
C
C     CONTRACTION ACCEPTED .
C
                  IF  ( NEXT ( NPL1 ) . LE . SIMP ( H ( NPL1 ) , NPL1 ) )  THEN
                      CALL  NEWVERTEX
C
C     IF STILL BAD SHRINK ALL VERTICES .
C
                  ELSE
                      DO  I = 1 , NPL1
                          DO  J = 1 , N
                              SIMP ( I , J ) = ( SIMP ( I , J ) + SIMP ( L ( NPL1 ) , J ) ) * BETA
                              TEMP ( J ) = SIMP ( I , J )
                          END  DO
                          TEMP ( NPL1 ) = 0D0
                          SIMP ( I , NPL1 ) = SUMPOWRES ( TEMP , NPL1 )
                      END  DO
                  END  IF
              END  IF
          END  IF
          CALL  RANK
C
C     CHECK FOR CONVERGENCE .
C
          DONE = . TRUE .
          DO  J = 1 , NPL1
              IF  (ABS ( SIMP ( H ( J ) , J ) ) . GT . TOLER . AND . ABS ( SIMP ( L ( J ) , J ) ) . GT .
     1        TOLER )  THEN
                  ERROR ( J ) = ABS ( ( SIMP ( H ( J ) , J ) - SIMP ( L ( J ) , J ) ) / SIMP ( H ( J ) , J ) )
              ELSE
```

Figure 7.5 (*continued*)

```
                    ERROR ( J ) = ABS ( SIMP ( H ( J ) , J ) - SIMP ( L ( J ) , J ) )
                    END  IF
                    IF  ( ERROR ( J ) . GT . TOLER )  DONE  = . FALSE .
                END  DO
                AMAX = 0D0
                DO  J = 1 , NPL1
                    IF  ( ABS ( ERROR ( J ) ) . GT . AMAX )  AMAX = ERROR ( J )
                END  DO
C               TYPE * , ' THE MAXIMUM ERROR IS : ' , AMAX
                NITER = NITER + 1
C               TYPE * . NITER , ' ITERATIONS HAVE BEEN PREFORMED '
                IF  ( DONE . EQ . FALSE . . AND . NITER . LT . LIM )  GOTO 10
                IF  ( NITER . GE . LIM )  STOP  ' TOO MANY ITERATIONS '
C
C     AVERAGE THE FINAL PARAMETERS .
C
                DO  I = 1 , NPL1
                    MEAN ( I ) = 0D0
                    DO  J = 1 , NPL1
                        MEAN ( I ) = MEAN ( I ) + SIMP ( J , I )
                    END  DO
                    MEAN ( I ) = MEAN ( I ) / NPL1
                END  DO
                TYPE * , NITER , ' ITERATIONS WERE PERFORMED '
C
C     WRITE THE FINAL OUTPUT PARAMETER VALUES .
C
                WRITE ( * , 15 ) ( MEAN ( I ) , I = 1 , N ) , MEAN ( NPL1 )
                IF ( M . EQ . N . OR . ABS ( POW - 2.0 ) . GT . 1D - 15 )  STOP  ' END OF PROCESSING '
                SIGMA = SQRT ( MEAN ( NPL1 ) / DFLOAT ( M - N ) )
15              FORMAT ( '        THE FINAL PARAMETER VALUES ARE : ' // < N > ( 7 X ,
     1          D22.15 / ) // ' THE SUM OF P - TH POWER OF THE RESIDUALS IS : ' // 7 X ,
     2          D22.15 )
C
C     CALCULATE PARTIAL DERIVATIVES NEEDED FOR COVARIANCE MATRIX .
C
                DO  I = 1 , N
                    DO  J = I , N
                        A ( J * ( J - 1 ) / 2 + I ) = 0.0
                    END  DO
                END  DO
                DO  K = 1 , M
                    READ ( LUNIT , 20 )  T , D
20                  FORMAT ( D22.15 )
                    DERV ( 1 ) = EXP ( X ( 2 ) * T )
                    DERV ( 2 ) = X ( 1 ) * EXP ( X ( 2 ) * T ) * T
                    DO  I = 1 , N
                        DO  J = I , N
                            A ( J * ( J - 1 ) / 2 + I ) = A ( J * ( J - 1 ) / 2 + I ) + DERV ( I ) * DERV ( J )
                        END  DO
                    END  DO
```

Figure 7.5 (*continued*)

```
            END  DO
            CLOSE ( LUNIT )
            CALL NORSLVE ( A , N )
            END
C
C     FUNCTION FOR COMPUTING THE SUM OF THE SELECTED POWER OF THE RESIDUALS .
C
            FUNCTION  SUMPOWRES ( X , N )
            REAL * 8  X ( N ) , SUMPOWRES , POW
            REAL * 16  F
            COMMON  / B /  M , POW , LUNIT
            F = 0.0
            DO I = 1 , M
              READ ( LUNIT , 5 )  T , D
5             FORMAT ( D22.15 )
              F = F + ABS ( ( X ( 1 ) * EXP ( T * X ( 2 ) ) - D ) ) ** POW
            END DO
            SUMPOWRES = F
            REWIND ( LUNIT )
            RETURN
            END
C
C     SUBROUTINE FOR RANKING THE VERTICES .
C
            SUBROUTINE  RANK
            REAL * 8  SIMP ( 100 , 100 ) , NEXT ( 100 )
            INTEGER  H ( 100 ) , L ( 100 )
            COMMON  / A /  SIMP , H , L , NEXT , NPL1
            DO  J = 1 , NPL1
              DO  I = 1 , NPL1
                IF  ( SIMP ( I , J ) . LT . SIMP ( L ( J ) , J ) )  L ( J ) = I
                IF  ( SIMP ( I , J ) . GT . SIMP ( H ( J ) , J ) )  H ( J ) = I
              END  DO
            END  DO
            RETURN
            END
C
C     SUBROUTINE FOR SELECTING A NEW VERTEX .
C
            SUBROUTINE  NEWVERTEX
            REAL * 8  SIMP ( 100 , 100 ) , NEXT ( 100 )
            INTEGER  H ( 100 ) , L ( 100 )
            COMMON  / A /  SIMP , H , L , NEXT , NPL1
            DO  I = 1 , NPL1
              SIMP ( H ( NPL1 ) , I ) = NEXT ( 1 )
            END DO
            RETURN
            END
```

Figure 7.5 (*continued*)

References

Abdelmalek, N.N. (1971). Linear L_1 Approximation for a Discrete Point Set and L_1 Solutions of Overdetermined Linear Equations, *J. ACM*, **18**, p. 41.

Acton, F.S. (1970). *Numerical Methods that (Usually) Work* (Harper and Row, New York).

Brent, R.P. (1973). *Algorithms for Minimization without Derivatives* (Prentice-Hall, Englewood Cliffs, N.J.).

Caceci, M.S. and Cacheris, W.P. (1984). Fitting Curves to Data, *BYTE*, **9**, No. 6, p. 340.

Dromey, R.G. (1982). *How to Solve It by Computer* (Prentice-Hall, Englewood Cliffs, N.J.).

Forsythe, G., Malcolm, M.A., and Moler, C.B. (1977). *Computer Methods for Mathematical Computations* (Prentice-Hall, Englewood Cliffs, N.J.).

Hald, J. and Madsen, K. (1985). Combined *LP* and Quasi-Newton Methods for Nonlinear L_1 Optimization, *SIAM J. Numer. Anal.*, **22**, p. 68.

Intriligator, M.D. (1981). Mathematical Programming with Applications to Economics. In Arrow, K.J. and Intriligator, M.D. (eds.) *Handbook of Mathematical Economics*, Vol. I (North-Holland, Amsterdam).

Kennedy, W.J., and Gentle, J.E. (1980). *Statistical Computing* (Marcel Dekker, New York).

Levenberg, K. (1944). A Method for the Solution of Certain Non-Linear Problems in Least Squares, *Quart. Appl. Math.*, **2**, p. 164.

Marquardt, D.W. (1963). An Algorithm for Least-Squares Estimation of Nonlinear Parameters, *J. SIAM*, **11**, p. 431.

Nelder, J.A. and Mead, R. (1965). A Simplex Method for Function Minimization, *Computer J.*, **7**, p. 308.

Patterson, E.M. (1969). *Topology* (Interscience, New York).

Press, W.H., Flannery, B.P., Teukolsky, S.A., and Vetterling, W.T. (1986). *Numerical Recipes: The Art of Scientific Computing* (Cambridge University Press, Cambridge).

Rosenbrock, H.H. (1960). An Automatic Method for Finding the Greatest or Least Value of a Function, *Computer J.*, **3**, p. 175.

Scheid, F. (1968). *Theory and Problems of Numerical Analysis* (McGraw-Hill, New York).

CHAPTER 8

The Singular Value Decomposition

8.1. Introduction

The singular value decomposition (SVD), closely related to matrix eigenvalue–eigenvector decompositions, is a powerful tool for analyzing linear systems. Like all mathematical tools it has its legitimate uses, but it can also be abused, of which we will have more to say in Section 8.4.

Recall Eq. (5.12). Any matrix $\mathbf{A}$ of size $m \times n$, with $m \geq n$, admits of the decomposition

$$\mathbf{A} = \mathbf{U} \cdot \mathbf{S} \cdot \mathbf{V}^\mathrm{T}, \tag{8.1}$$

where $\mathbf{U}$ is an $m \times m$ orthogonal matrix, $\mathbf{V}$ is an $n \times n$ orthogonal matrix, and $\mathbf{S}$ is an $m \times n$ matrix whose lower $(m - n) \times n$ part is null and whose upper $n \times n$ part is diagonal. The diagonal entries of $\mathbf{S}$ are the singular values of $\mathbf{A}$. When $\mathbf{A}$ is of full rank, n for Eq. (8.1), the n singular values are nonzero. If $\mathbf{A}$ is subrank, say rank r $(r < n)$, then $n - r$ of the singular values are zero. (The restriction $m \geq n$ is unnecessary because the SVD can be applied even to underdetermined systems by adding sufficient null rows to make $m = n$; but our interest in this book is overdetermined systems.)

Multiply Eq. (8.1) on the left by $\mathbf{A}^\mathrm{T}$. We find that

$$\mathbf{A}^\mathrm{T} \cdot \mathbf{A} = \mathbf{V} \cdot \mathbf{S}^2 \cdot \mathbf{V}^\mathrm{T}; \tag{8.2}$$

the right-hand side is an eigenvalue–eigenvector decomposition for $\mathbf{A}^\mathrm{T} \cdot \mathbf{A}$ and coincides with Eq. (7.47), which explains the use of $\mathbf{S}^2$ rather than $\mathbf{S}$ in that equation. Multiply Eq. (8.1) on the right by $\mathbf{A}^\mathrm{T}$. Then

$$\mathbf{A} \cdot \mathbf{A}^\mathrm{T} = \mathbf{U} \cdot \mathbf{S}^2 \cdot \mathbf{U}^\mathrm{T}. \tag{8.3}$$

In Eq. (8.2), $\mathbf{S}^2$ is $n \times n$ and in Eq. (8.3) is $m \times m$, but in the latter the lower $m - n$ eigenvalues are zero, and the upper n eigenvalues coincide with those of Eq. (8.2); the $\mathbf{S}^2$ of Eq. (8.3) is the $\mathbf{S}^2$ of Eq. (8.2) augmented by null rows and columns.

$\mathbf{A}^\mathrm{T} \cdot \mathbf{A}$ and $\mathbf{A} \cdot \mathbf{A}^\mathrm{T}$ have the same eigenvalues, the squares of the singular values of $\mathbf{A}$. The columns of $\mathbf{V}$ are the eigenvectors of $\mathbf{A}^\mathrm{T} \cdot \mathbf{A}$ and the columns of $\mathbf{U}$ are the eigenvectors of $\mathbf{A} \cdot \mathbf{A}^T$. Recalling the discussion of the condition number of a matrix in Chapter 2 and remembering Eq. (2.28), we can define the condition number of $\mathbf{A}$ as

$$\text{COND}(\mathbf{A}) = \frac{s_{\max}}{s_{\min}}, \tag{8.4}$$

where the s's are the singular values.

Singular values, as Forsythe, Malcolm, and Moler (1977, pp. 206–207) show, have a geometrical interpretation. Consider the L_2 norm—in this section all norms are L_2 unless stated otherwise—of the product $\mathbf{A} \cdot \mathbf{X}$,

$$\|\mathbf{A} \cdot \mathbf{X}\| = \|\mathbf{U} \cdot \mathbf{S} \cdot \mathbf{V}^T \cdot \mathbf{X}\|. \tag{8.5}$$

Let $\mathbf{y} = \mathbf{V}^T \cdot \mathbf{X}$. Because orthogonal matrices conserve the L_2 norm (see Eq. (5.48))

$$\|\mathbf{y}\| = \|\mathbf{X}\|. \tag{8.6}$$

Then

$$\|\mathbf{A} \cdot \mathbf{X}\| = \|\mathbf{U} \cdot \mathbf{S} \cdot \mathbf{y}\| = \|\mathbf{S} \cdot \mathbf{y}\|. \tag{8.7}$$

But $\mathbf{S}$ is a diagonal matrix. Therefore,

$$s_{\min} \|\mathbf{y}\| \leq \|\mathbf{S} \cdot \mathbf{y}\| \leq s_{\max} \|\mathbf{y}\|, \tag{8.8}$$

or

$$s_{\min} \leq \frac{\|\mathbf{A} \cdot \mathbf{X}\|}{\|\mathbf{X}\|} \leq s_{\max}. \tag{8.9}$$

Consider the special case $\|\mathbf{X}\| = 1$ and $n = 2$. The norm of $\mathbf{X}$ traces out a circle. The norm $\mathbf{A} \cdot \mathbf{X}$ transforms the circle into a new figure. When $\mathbf{A}$ is orthogonal the new figure is also a circle because not only do orthogonal matrices conserve the L_2 norm, they also have unit condition number with $s_{\max} = s_{\min} = 1$. But if $\mathbf{A}$ is other than orthogonal, Eq. (8.9) shows that the circle is transformed into an ellipse whose two axes are $s_{\max}$ and $s_{\min}$. Figure 8.1 illustrates the situation.

As the condition number Eq. (8.4) increases, the ellipse becomes more and more elongated. For a singular matrix, $s_{\min} = 0$, the ellipse degenerates into a straight-line segment. Whereas the components of $\mathbf{X}$ are uncorrelated by hypothesis, the components of $\mathbf{A} \cdot \mathbf{X}$ are correlated, the correlation becoming stronger as the eccentricity of the ellipse increases until, for a singular matrix, it becomes perfect.

To illustrate the SVD take the matrix

$$\mathbf{A} = \begin{pmatrix} 1 & 1 & 1 \\ 1 & 2 & 4 \\ 1 & 3 & 9 \\ 1 & 4 & 16 \\ 1 & 5 & 25 \end{pmatrix}. \tag{8.10}$$

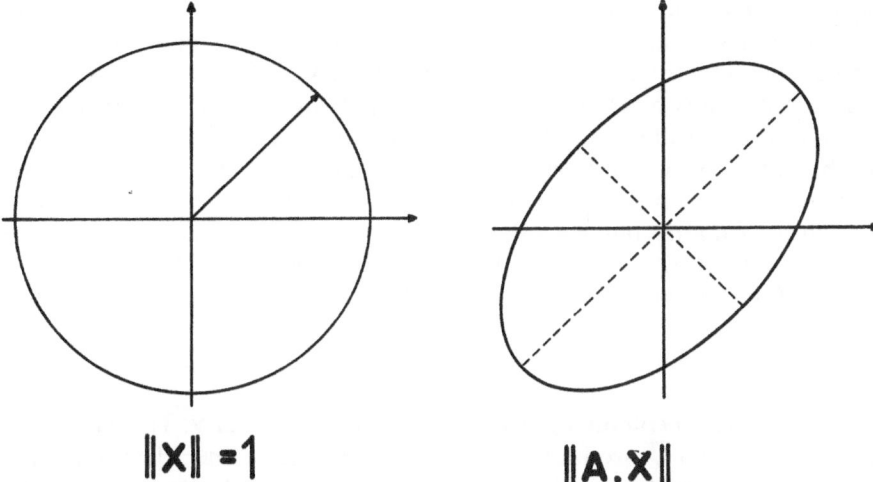

Figure 8.1. Distortion of unit circle by transformation $A \cdot X$.

Matrix (8.10) decomposes into, to five decimals,

$$U = \begin{pmatrix} -0.03895 & 0.52790 & 0.77815 & -0.00891 & -0.33794 \\ -0.13670 & 0.58904 & -0.07600 & 0.24357 & 0.75448 \\ -0.29496 & 0.45745 & -0.43526 & -0.67727 & -0.23578 \\ -0.51373 & 0.13315 & -0.29963 & 0.65945 & -0.44011 \\ -0.79301 & -0.38388 & 0.33088 & -0.21685 & 0.25935 \end{pmatrix},$$

$$S = \begin{pmatrix} 32.15633 & 0 & 0 \\ 0 & 2.19773 & 0 \\ 0 & 0 & 0.37438 \\ 0 & 0 & 0 \\ 0 & 0 & 0 \end{pmatrix},$$

and

$$V = \begin{pmatrix} -0.05527 & 0.60229 & 0.79636 \\ -0.22444 & 0.76968 & -0.59768 \\ -0.97292 & -0.21177 & 0.09264 \end{pmatrix}.$$

The next section explains how the decomposition was obtained.

The condition number of the matrix is 86. The norm $\|X\| = 1$, tracing a sphere in three dimensions, becomes distorted to an ellipsoid with axes given by the singular values.

Because the lower $(m - n) \times n$ part of S is null in Eq. (8.1), only n columns of U—fewer if A is rank deficient—contribute to the triple matrix product on the right-hand side. We could realize a saving of space by storing only the significant n columns of $\mathbf{U}$, even though the resulting $m \times n$ matrix $\mathbf{U}$ loses its orthogonality because it is no longer square. But its columns are still orthogonal. For the example just given of the SVD of the matrix (8.10), the last two columns of $\mathbf{U}$ could be omitted and $\mathbf{S}$ expressed as a true $n \times n$ diagonal matrix without affecting our ability to reconstruct $\mathbf{A}$ from the matrices $\mathbf{U}$, $\mathbf{S}$, and $\mathbf{V}$. Some authorities, such as Press, Flannery, Teukolsky, and Vetterling (1986, pp. 52–53), define the SVD taking $\mathbf{U}$ as an $m \times n$ matrix and $\mathbf{S}$ as an $n \times n$ matrix; but use of the truncated $\mathbf{U}$ and $\mathbf{S}$ matrices really repesents a special case of the more general decomposition defined by Eq. (8.1).

Use of the truncated version of the SVD permits expression of Eq. (8.1) as

$$\mathbf{A} = \sum_{i=1}^{n} s_i \mathbf{u}_i \cdot \mathbf{v}_i^T, \tag{8.11}$$

where $\mathbf{u}_i$ is the ith column of $\mathbf{U}$ and $\mathbf{v}_i$ is the ith column of $\mathbf{V}$. The backwards product of $\mathbf{u}_i$ and $\mathbf{v}_i^T$ produces an $m \times n$ matrix of rank one. The summation of n of these backwards products, each product weighted by its corresponding singular value, produces the rank n matrix $\mathbf{A}$. If $\mathbf{A}$ is subrank the summation stops at the upper index r, the rank of $\mathbf{A}$, instead of the upper index n. Eq. (8.11) reveals more clearly than Eq. (8.1) the nature of subrank matrices. Each backwards product is rank one because the columns are multiples of one another. Summing n backwards products, produces n independent columns in $\mathbf{A}$. Even if some of the s_i are the same, their corresponding $\mathbf{u}_i$ and $\mathbf{v}_i$ are not—think of the eigenvectors of the unit matrix. But stopping the summation before reaching n means that some of the columns of $\mathbf{A}$ must be dependent and hence $\mathbf{A}$ is subrank. Notice the affinity between Eqns. (8.1) and (8.11) and Eqns. (7.47) and (7.48).

8.2. Calculating the SVD

Equations (8.2) and (8.3), and the defining equations for matrix eigenvalues and eigenvectors (Eqns (2.29) and (2.30)), furnish a direct, but impractical, way to calculate the SVD. From the matrix $\mathbf{A}^T \cdot \mathbf{A}$, find its characteristic equation by the transformation of Eq. (2.30) into a polynomial, and calculate all the roots of the polynomial. The roots are the squares of the singular values. Then solve n homogeneous linear systems to obtain $\mathbf{V}$. The same procedure produces $\mathbf{U}$ from Eq. (8.3). But far too much work is involved in such a simple-minded approach and, even more important, too much chance of gross errors in the result. The roots of a high-order polynomial can be extremely sensitive to minute changes in the coefficients. Vandergraft (1983, pp. 303–304) analyzes a horrifying example often encountered in the literature, Wilkinson's polynomial of order twenty. A change in one of the coefficients by an amount 2^{-23}, the twenty-third bit in the mantissa of a floating-point number, produces enormous changes in some of the roots. The root 20.00000 becomes 20.84691 and ten of the total of twenty real roots become complex with significant imaginary parts.

Rather than use the defining equations directly, we could apply eigenvalue–eigenvector techniques to the matrices $\mathbf{A}^T \cdot \mathbf{A}$ and $\mathbf{A} \cdot \mathbf{A}^T$. Although this book is not specifically concerned with matrix eigenvalues, given the close relationship between singular values and eigenvalues, a brief discussion of practical eigenvalue–eigenvector techniques seems warranted.

Similarity transformations constitute the basic mathematical tool for practical eigenvalue routines. Let $\mathbf{A}$ and $\mathbf{B}$ be square matrices with real elements, and let $\mathbf{T}$ be any nonsingular square matrix with the same dimensions as $\mathbf{A}$ and $\mathbf{B}$. A similarity transformation is defined by

$$\mathbf{B} = \mathbf{T} \cdot \mathbf{A} \cdot \mathbf{T}^{-1}. \tag{8.12}$$

The importance of similarity transformations arises from their conserving eigenvalues. If λ is an eigenvalue of $\mathbf{A}$,

$$\mathbf{A} \cdot \mathbf{X} = \lambda \mathbf{X},$$

then

$$\mathbf{T} \cdot \mathbf{A} \cdot \mathbf{X} = \lambda \mathbf{T} \cdot \mathbf{X}.$$

Let

$$\mathbf{y} = \mathbf{T} \cdot \mathbf{X}. \tag{8.13}$$

Then

$$\mathbf{T} \cdot \mathbf{A} \cdot \mathbf{T}^{-1} \cdot \mathbf{y} = \lambda \mathbf{y},$$

or

$$\mathbf{B} \cdot \mathbf{y} = \lambda \mathbf{y}.$$

Both $\mathbf{A}$ and $\mathbf{B}$ have the same eigenvalue λ, and the eigenvector $\mathbf{y}$ of $\mathbf{B}$ is related to the eigenvector $\mathbf{X}$ of $\mathbf{A}$ by Eq. (8.13).

Equation (8.12) would hardly be practical if we had to invert $\mathbf{T}$ to calculate the transformation. But if $\mathbf{T}$ is selected as an orthogonal matrix its transpose is its inverse, and Eq. (8.12) requires only two matrix multiplications. We see once again the great utility of orthogonal matrices. They conserve the Euclidean norm of vectors, of use in least squares problems, and permit the easy calculation of similarity transformations for use in eigenvalue–eigenvector problems.

Jacoby's method, applicable only to symmetric matrices, takes for $\mathbf{T}$ a Givens rotation, Eq. (5.59), or reflection, Eq. (5.73), matrix. θ is selected to zero a superdiagonal element and its symmetric subdiagonal companion. Unfortunately, as new elements are zeroed nonzeros reappear in elements already annihilated. But after a complete sweep over all off-diagonal elements their sum of squares decreases. After a theoretically infinite number of iterations all off-diagonal elements are zero and we have the matrix $\mathbf{S}^2$. The same transformations applied to the unit matrix reduce it to the matrix of the eigenvectors. In practice about $6n^3$ operations are required to effect an eigenvalue–eigenvector decomposition by Jacoby's method. For details, consult Acton (1970, pp. 319–322).

The QR method does better. Wilkinson (1965, Chap. 8) is the best reference for this method. Rather than take a symmetric matrix all the way to diagonal form in a theoretically infinite number of steps, the QR method first reduces it to a tridiagonal matrix by the application of a finite number of Householder transformations (Eq. (5.68)). Then an iterative phase reduces the tridiagonal matrix to diagonal by use of the Givens transformations. Although the iterative phase is also, as with Jacoby's method, theoretically infinite, in practice, as soon as the off-diagonal elements are of the order of the machine epsilon, the method terminates. An ancillary algorithm calculates the eigenvectors. Because the iterative phase of the method is applied to a tridiagonal, rather than a full, matrix, the QR method is more efficient than Jacoby's. About $2n^3/3 + 44n^2$ operations effect an eigenvalue–eigenvector decomposition of a symmetric matrix.

For the efficient execution performance of an algorithm based on the QR method, origin shifts are necessary. We introduced shifting in Section 2.3, in connection with finding the smallest eigenvalue of a matrix by the power method. If λ is an eigenvalue of $\mathbf{A}$ and we subtract an origin shift k from it we find

$$\mathbf{A} \cdot \mathbf{X} = \lambda \mathbf{X} = (\lambda - k)\mathbf{X} + k\mathbf{X}$$

or

$$(\mathbf{A} - k\mathbf{I}) \cdot \mathbf{X} = (\lambda - k) \cdot \mathbf{X}. \tag{8.14}$$

The eigenvector remains the same. The eigenvalue solved for, $\lambda - k$, permits recovery of λ by adding the shift parameter previously subtracted off.

Origin shifts become important, because we can prove that the off-diagonal elements in the iterative phase of the QR method go to zero proportional to $(\lambda_i/\lambda_j)^p$, where i and j are the index positions of the element A_{ij} and p, is a parameter greater than unity that increases with each iteration; see Wilkinson (1965, Sec. 8.29) for details. The eigenvalues can be ordered so that $\lambda_i < \lambda_j$. Thus, $(\lambda_i/\lambda_j)^p$ always converges to zero. But if λ_i is close to λ_j in absolute value, convergence will be slow. By careful selection of the shift parameter k the rate of convergence $(\lambda_i - k)^p/(\lambda_j - k)^p$ can be markedly accelerated. We will see shortly how k is determined.

The QR method is more efficient than Jacoby's method for matrix eigenvalue–eigenvector problems. We would, nevertheless, have to apply it twice, once to the symmetric matrix $\mathbf{A}^T \cdot \mathbf{A}$ and once to the symmetric matrix $\mathbf{A} \cdot \mathbf{A}^T$. The eigenvalues would be calculated twice, which is hardly desirable or efficient. Furthermore, the condition numbers of $\mathbf{A}^T \cdot \mathbf{A}$ and $\mathbf{A} \cdot \mathbf{A}^T$ are the square of the condition number of $\mathbf{A}$, and we can expect at least some loss of accuracy in the results as a consequence.

Fortunately, the QR method can be modified and applied directly to $\mathbf{A}$ to effect the decomposition, Eq. (8.1). The following discussion of the SVD as implemented by a modified QR method is based on Golub and Reinsch's (1970) paper, which includes a program in ALGOL. This program, or modifications of it, translated into FORTRAN has found its way into numerous sources, such as Lawson and Hanson (1974, pp. 298–300), Forsythe, Malcolm, and Moler (1977, pp. 229–235), and Press, Flannery, Teukolsky, and Vetterling (1986, pp. 60–64). Because the original article and the references just given contain complete programs, one will not be given here.

The Golub and Reinsch QR algorithm—algorithm, not method, because we are speaking about a specific implementation—proceeds in two stages. The first stage reduces the $m \times n$ matrix $\mathbf{A}$ to upper bidiagonal form by a series of $2n - 2$ Householder transformations,

$$\mathbf{B} = \mathbf{L}_n \cdot \mathbf{L}_{n-1} \cdots \mathbf{L}_1 \cdot \mathbf{A} \cdot \mathbf{R}_1 \cdot \mathbf{R}_2 \cdots \mathbf{R}_{n-2}. \tag{8.15}$$

$\mathbf{L}_i$ is a Householder transformation applied on the left to zero the elements $i + 1$ through m of column i. The Householder transformation $\mathbf{R}_i$ applied on the right zeroes elements $i + 2$ through n of row i. The transformations are applied in the order $\mathbf{L}_1$, $\mathbf{R}_1$, $\mathbf{L}_2$, $\mathbf{R}_2$, ..., $\mathbf{L}_n$. $\mathbf{B}$ is a matrix whose lower $(m - n) \times n$ part is null and whose upper $n \times n$ part contains a diagonal and one superdiagonal, hence the name bidiagonal.

Both the matrix $\mathbf{A}$ and the matrix $\mathbf{B}$ have the same singular values. Let $\mathbf{L} = \mathbf{L}_n \cdot \mathbf{L}_{n-1} \cdots \mathbf{L}_1$ and $\mathbf{R} = \mathbf{R}_1 \cdot \mathbf{R}_2 \cdots \mathbf{R}_{n-2}$. Then

$$\mathbf{B} = \mathbf{L} \cdot \mathbf{A} \cdot \mathbf{R} = \mathbf{L} \cdot \mathbf{U} \cdot \mathbf{S} \cdot \mathbf{V}^T \cdot \mathbf{R} = \mathbf{U}' \cdot \mathbf{S} \cdot \mathbf{V}'^T, \tag{8.16}$$

with $\mathbf{U}' = \mathbf{L} \cdot \mathbf{U}$ and $\mathbf{V}' = \mathbf{R}^T \cdot \mathbf{V}$. Thus, $\mathbf{B}$ has the same singular values $\mathbf{S}$ as those of $\mathbf{A}$ and left $\mathbf{U}'$ and right $\mathbf{V}'$ orthogonal matrices of singular vectors— named by analogy with the terms "eigenvalue" and "eigenvector"—related to the $\mathbf{U}$ and $\mathbf{V}$ of $\mathbf{A}$ by Eq. (8.16).

Before proceeding to the second stage of the SVD algorithm we will illustrate the concepts discussed so far with the bidiagonalization of the matrix (8.10). The first Householder transformation $\mathbf{L}_1$ zeroes all of the elements of the first column except A_{11}. This transformation has already been calculated as the first step in the triangularization of Eq. (5.64) and is given as the 5×5 matrix $\mathbf{H}$ following Eq. (5.76). Application of this transformation to (8.10) results in

$$\begin{pmatrix} -2.23607 & -6.70820 & -24.59675 \\ 0 & -0.38197 & -3.90983 \\ 0 & 0.61803 & 1.09017 \\ 0 & 1.61803 & 8.09017 \\ 0 & 2.61803 & 17.09017 \end{pmatrix}.$$

The matrix $\mathbf{R}_1$ to zero the element -24.59675 is calculated next. Because we wish to annihilate a single element a 3×3 Givens transformation is all that is needed. The matrix is

$$\mathbf{R}_1 = \begin{pmatrix} 1 & 0 & 0 \\ 0 & -0.26312 & -0.96476 \\ 0 & -0.96476 & 0.26312 \end{pmatrix}.$$

For larger matrices $\mathbf{A}$, of course, we would construct a Householder transformation to zero a portion of an entire row, not just a single element.

Upon postmultiplying $\mathbf{L}_1 \cdot \mathbf{A}$ by $\mathbf{R}_1$ we obtain

$$
\begin{pmatrix}
-2.23607 & 25.49510 & 0 \\
0 & 3.87256 & -0.66024 \\
0 & -1.21437 & -0.30941 \\
0 & -8.23084 & 0.56764 \\
0 & -17.17683 & 1.97094
\end{pmatrix}.
$$

The application of two more Householder transformations on the left—no more right Givens transformations are needed—results in the upper bidiagonal matrix $\mathbf{B}$,

$$
\begin{pmatrix}
-2.23607 & 25.49510 & 0 \\
0 & -19.47464 & 2.09029 \\
0 & 0 & 0.60757 \\
0 & 0 & 0 \\
0 & 0 & 0
\end{pmatrix}.
$$

Although for this example we calculated explicit Householder and Givens transformation matrices, in practice we would not use an explicit matrix representation of the transformation, but would rather apply the Householder transformations in factored form and the Givens transformations only to the elements actually modified, as explained in Section 5.4.

The second stage of the SVD algorithm is somewhat more complicated. The bidiagonal matrix $\mathbf{B}$ cannot be reduced to the diagonal matrix $\mathbf{S}$ in a finite number of steps. Application of a right Givens transformation to zero a superdiagonal element indeed annihilates the element, but also introduces a nonzero subdiagonal element on the diagonal just below the main diagonal. Application of a left Givens transformation rezeroes the subdiagonal element, but reintroduces a superdiagonal element. But after a complete sweep over all superdiagonal elements, their sum of squares decreases. We continue the process of applying right and left Givens transformations until the superdiagonal elements are of the order of the machine epsilon, the same idea encountered in Jacoby's method.

The SVD of the bidiagonal matrix $\mathbf{B}$ proceeds by a series of iterative steps, the rationale for which is far from obvious. This section presents them in cookbook fashion. The reader should consult Golub and Reinsch's original article (1970) and Lawson and Hanson (1974, Chap. 18) for details.

Let $\mathbf{B}_i$ be a series of matrix iterates that start with $\mathbf{B}_i = \mathbf{B}$ and converge to the diagonal matrix of singular values $\mathbf{S}$. Likewise, $\mathbf{V}'_i$ and $\mathbf{U}'_i$ are orthogonal matrix iterates, $\mathbf{V}'_i$ is a Givens transformation applied on the right, and $\mathbf{U}'_i$ is a Givens transformation applied on the left, whose products converge to the $\mathbf{V}'$ and $\mathbf{U}'$ of Eq. (8.16), $\mathbf{V}' = \mathbf{V}'_1 \cdot \mathbf{V}'_2, \ldots, \mathbf{U}' = \cdots \mathbf{U}'_2 \cdot \mathbf{U}'_1$.

The first right Givens transformation is arbitrary. For correct functioning of the SVD algorithm V_1' should be constructed to zero the subdiagonal element in the first column of

$$V_1'^T \cdot (B^T \cdot B - kI), \tag{8.17}$$

where k is a shift parameter defined below. Because B is bidiagonal, $B^T \cdot B$ is tridiagonal and only one subdiagonal element in each column is nonzero. Subsequent iterations construct V_i so that

$$V_i'^T \cdot (B_i^T \cdot B_i - k_i I) \tag{8.18}$$

is upper triangular. The k_ith shift parameter is taken as the eigenvalue of the lower right 2×2 submatrix of $B_i^T \cdot B_i$ closest to the lower right element of $B_i^T \cdot B_i$. Wilkinson (1965, Sec. 8.36) explains why such a seemingly weird selection of origin shift is effective. Many hours of number crunching with the computer corroborates this strategy for origin shift selection. The matrix U_i' is constructed so that

$$B_{i+1} = U_i'^T \cdot B_i \cdot V_i' \tag{8.19}$$

is upper bidiagonal.

To illustrate these ideas we continue with the SVD of the matrix (8.10). From the bidiagonal matrix already calculated we find

$$B^T \cdot B = \begin{pmatrix} 5.00000 & -57.00877 & 0 \\ -57.00877 & 1029.26149 & -40.70769 \\ 0 & -40.70769 & 4.73846 \end{pmatrix}.$$

The shift parameter k of Eq. (8.17) comes from the eigenvalues of the lower right 2×2 submatrix of $B^T \cdot B$

$$\begin{vmatrix} 1029.26149 - \lambda & -40.70769 \\ -40.70769 & 4.73846 - \lambda \end{vmatrix} = 0.$$

This characteristic equation has two eigenvalues, $\lambda_1 = 1030.87640$ and $\lambda_2 = 3.12356$. λ_2 is closest to the lower right element 4.73846 of $B^T \cdot B$ and, hence, k is taken as λ_2.

V_1' is constructed as the Givens matrix to zero the element -57.00877 in the first column of

$$B^T \cdot B - kI = \begin{pmatrix} 1.87644 & -57.00877 & 0 \\ -57.00877 & 1026.13794 & -40.70769 \\ 0 & -40.70769 & 1.61491 \end{pmatrix}.$$

The required matrix is

$$V_1'^T = \begin{pmatrix} 0.03290 & -0.99946 & 0 \\ -0.99946 & -0.03290 & 0 \\ 0 & 0 & 1 \end{pmatrix}.$$

Upon multiplying **B** by V'_1 we obtain

$$\mathbf{B}_1 \cdot \mathbf{V}'_1 = \begin{pmatrix} -25.55490 & 1.39607 & 0 \\ 19.46412 & 0.64072 & 2.09029 \\ 0 & 0 & 0.60757 \\ 0 & 0 & 0 \\ 0 & 0 & 0 \end{pmatrix}.$$

Notice that V'_1 annihilates no superdiagonal element; V'_1 is constructed to satisfy the condition (8.17). Subsequent V'_i's annihilate superdiagonal elements. We now need a U'_1 to zero the element 19.46412. The required matrix is

$$\mathbf{U}'^{\mathrm{T}}_1 = \begin{pmatrix} -0.79553 & 0.60592 & 0 & 0 & 0 \\ -0.60592 & -0.79553 & 0 & 0 & 0 \\ 0 & 0 & 1 & 0 & 0 \\ 0 & 0 & 0 & 1 & 0 \\ 0 & 0 & 0 & 0 & 1 \end{pmatrix}.$$

We find from the actual QR algorithm, whose details differ from the above as explained shortly,

$$\mathbf{B}_2 = \begin{pmatrix} 32.12323 & 1.45812 & 0 \\ 0 & 0.93574 & -1.82254 \\ 0 & 0 & 0.88018 \\ 0 & 0 & 0 \\ 0 & 0 & 0 \end{pmatrix}.$$

After the first pass through the iterative phase of the SVD algorithm the sum of the squares of the off-diagonal elements has decreased from 654.36944 to 5.44777. A total of five iterations reduces the off-diagonal elements to zero to five decimals. We have

$$\mathbf{B}_5 = \begin{pmatrix} 32.15633 & 0 & 0 \\ 0 & 2.19773 & 0 \\ 0 & 0 & 0.37438 \\ 0 & 0 & 0 \\ 0 & 0 & 0 \end{pmatrix}$$

and

$$\mathbf{V}'_5 = \begin{pmatrix} 0.05527 & 0.60229 & -0.79636 \\ -0.99769 & 0.00180 & -0.06789 \\ 0.03946 & -0.79828 & -0.60100 \end{pmatrix}.$$

U'_5 will not be shown. Thus, $\mathbf{B}_5$ has the same singular values as **A**. Premultiplication of V'_5 by $\mathbf{R}^{\mathrm{T}}_1$ reproduces **V** (with the signs of some columns scrambled; the SVD takes a rather cavalier attitude towards signs of columns of orthogonal matrices. As long as we can reproduce the matrix **A** from its SVD, signs of columns are unimportant).

Rapid convergence characterizes the iterative phase of the SVD algorithm. Lawson and Hanson (1974, App. B) prove that convergence is quadratic. What about the operation count? Performing an analysis similar to those in Chapter 5 for the operation counts for Cholesky decomposition and the Givens and Householder transformations on Golub and Reinsch's ALGOL code (1970), again keeping only cubic terms, we find that the first stage of the QR algorithm, the bidiagonalization of the $m \times n$ matrix A, requires $2mn^2 - 4n^3/3$ operations. To accumulate the $n \times n$ matrix R necessitates an additional $4n^3/3$ operations. The accumulation of L is expensive, but fortunately it is often not needed. With the truncated version of SVD L is $m \times n$ and needs $mn^2 - n^3/3$ operations to form; if L is the full $m \times m$ orthogonal matrix then $2m^2n - 2mn^2 + 2n^3/3$ operations are required.

The operation count for the iterative stage of the algorithm presents greater difficulties. It is conceivable, but hardly likely, that after the first stage the superdiagonal elements of the bidiagonal matrix B are already zero, and the second stage becomes unnecessary. If all of the superdiagonal elements are nonzero and large, the diagonalization of B needs, in the worst case, $2n^2$ operations plus the calculation of $n^2/2$ square roots per iteration. Calculation of V' requires, in the worst case, $2n^3$ operations and of U', $4mn^2$ operations per iteration. But the worst case hardly ever occurs in practice. The details of the SVD algorithm differ from Eqns (8.17)–(8.19) and the numerical example presented. The matrix $B^T \cdot B$ is never formed, origin shifts are performed implicitly, and as soon as a superdiagonal element becomes zero the matrix B_i is partitioned into submatrices. This decreases significantly the operation count. As a rough rule of thumb we can consider that calculation of V' requires, in practice, about $2n^2$ operations and of U' about $4mn$ operations; the operation count for the diagonalization of B remains the same. Lawson and Hanson (1974, Table 19.1) assert that $2n$ iterations suffice for convergence. Notice that calculation of U, particularly the full $m \times m$ matrix, represents a significant additional computational burden.

8.3. Total Least Squares

The SVD has many applications in numerical mathematics. One of the most useful is total least squares. The discussion in Chapter 5 of linear least squares assumed that all of the observational error in Eq. (4.1) is concentrated in the vector d of data. Section 4.3 explains that this assumption is usually defensible because the matrix A comes from a mathematical model and, even though the model may be incorrect, incorporates no observational error. There may be, nevertheless, occasions when A itself comes from observations or a model derived from observations and may not be assumed error-free. Total least squares calculates a least squares solution when both A and d contain errors.

But total least squares has nothing to say about A's being based on an inadequate model. **A**, whether or not it incorporates observational error, may inadequately model the physical situation under consideration. Press, Flannery, Teukolsky, and Vetterling (1986, pp. 502–503) feel that one should check for an inadequate model by calculating a chi-square goodness of fit statistic, defined by

$$\chi^2 = \left\| \frac{\mathbf{A} \cdot \mathbf{X} - \mathbf{d}}{\sigma} \right\|_2, \tag{8.20}$$

where σ is an m vector of the errors of each data point. The theoretical chi-square statistic, derived in Mathews and Walker (1964, pp. 368–369), assumes that the data are normally distributed. To the extent that this assumption is violated, because the data obey some other distribution or the model is incorrect, Eq. (8.20), calculated from the data, differs from its theoretical counterpart.

Unfortunately, Eq. (8.20) assumes that the errors σ_i of the data points are known, an assumption of often dubious validity, especially at the beginning of an experiment. Indeed, frequently, part of the function of analyzing data is to find a mean error of unit weight, $\sigma(1)$, for the data set as a whole. Nor does Eq. (8.20) have anything to say about **A** being corrupted by observational error. The author feels that an inspection of the residuals affords the best way to detect an inadequate model, although Eq. (8.20) may be useful if we in fact have the individual σ_i's.

Two concepts from linear algebra are of utility in connection with total least squares, the null space and range of a matrix. Given a square matrix **A** its null space is the set of vectors **X** for which $\mathbf{A} \cdot \mathbf{X} = \mathbf{0}$. Its range is the set of vectors **b** for which $\mathbf{A} \cdot \mathbf{X} = \mathbf{b}$ has a solution **X**. If **A** is full rank these concepts are of little use because the null space of **A** consists only of the vector $\mathbf{X} = \mathbf{0}$, and, because $\mathbf{X} = \mathbf{A}^{-1} \cdot \mathbf{b}$, all **b** are in the range of **A**. But if **A** is subrank the concepts have more meaning.

Consider the SVD of an $m \times n$ matrix **A**,

$$\mathbf{A} \cdot \mathbf{V} = \mathbf{U} \cdot \mathbf{S} \quad \Rightarrow$$

$$\mathbf{A} \cdot (\mathbf{v}_1 \quad \mathbf{v}_2 \quad \cdots \quad \mathbf{v}_n) = (\mathbf{u}_1 \quad \mathbf{u}_2 \quad \cdots \quad \mathbf{u}_m) \cdot \begin{pmatrix} s_1 & 0 & \cdots & 0 & \cdots & 0 \\ 0 & s_2 & & 0 & & 0 \\ \vdots & & & \vdots & & \vdots \\ & & & s_n & & \\ & & & \vdots & & \\ 0 & \cdots & & 0 & \cdots & 0 \end{pmatrix}, \tag{8.21}$$

where the $\mathbf{v}_i$'s are the columns of **V** and the $\mathbf{u}_i$'s are the columns of **U**. If an s_i is zero $\mathbf{A} \cdot \mathbf{v}_i = \mathbf{0}$, and $\mathbf{v}_i$ is in the null space of **A**. The number of columns of **V** for which $\mathbf{A} \cdot \mathbf{v}_i = \mathbf{0}$ determines the rank of **A**, the row dimension of **A** minus the number of null singular values. The SVD thus provides a practical way of finding the rank of a matrix, more practical than the use of the number of linearly independent columns of **A** in Chapter 2, a concept intuitively evident but difficult to implement.

If s_i is other than zero, $\mathbf{u}_i$ is in the range of **A**. The set of vectors $\mathbf{u}_i$, for which s_i is not null, forms an orthogonal basis for the range of **A**. Any vector in the range can be constructed as a linear combination of the $\mathbf{u}_i$'s.

Notice that Eq. (8.21) involves an $m \times n$ matrix **A**. Thus, the concepts of null space and range are also applicable to rectangular, not just square, matrices.

To illustrate the idea of null space and range, take as a trivial example the SVD of the singular matrix

$$\mathbf{A} = \begin{pmatrix} 1 & 0 \\ 0 & 0 \end{pmatrix}.$$

The SVD gives

$$\mathbf{S} = \begin{pmatrix} 1 & 0 \\ 0 & 0 \end{pmatrix}, \quad \mathbf{U} = \begin{pmatrix} 1 & 0 \\ 0 & 1 \end{pmatrix}, \quad \mathbf{V} = \begin{pmatrix} 1 & 0 \\ 0 & 1 \end{pmatrix}.$$

The second column of **V**, $\mathbf{v}_2^T = (0\ \ 1)$, forms a basis for the null space of **A** because $\mathbf{A} \cdot \mathbf{v}_2 = \mathbf{0}$. The first column of **U**, $\mathbf{u}_1^T = (1\ \ 0)$, constitutes a basis for the range of **A**. In a standard two-dimensional coordinate system $X - y$ any vector lying along the X-axis is in the range of **A**. But such a simple vector as $\mathbf{b}^T = (1\ \ 1)$ falls outside the range of **A**. The reason is manifest. **A** is singular and maps any two-dimensional vector into a one-dimensional vector. Thus, unless the y component of a two-dimensional happens to be zero, generally not the case, a two-dimensional vector lies outside **A**'s range. This is true in general. When **A** is subrank, say rank $n - r$, a vector of dimension higher than $n - r$ falls outside the range of **A**.

What has all of this to do with total least squares? A careful analysis of the total least squares problem, by Golub and Van Loan (1980), upon which our discussion will be based, starts with the basic least squares problem, minimize $\mathbf{r}^T \cdot \mathbf{r}$ with $\mathbf{r}$ given by Eq. (4.2), viewed somewhat differently. We have

$$\mathbf{A} \cdot \mathbf{X} = \mathbf{d} + \mathbf{r}. \tag{8.22}$$

$\mathbf{d} + \mathbf{r}$ must be in the range of **A**, and we seek the vector $\mathbf{r}$ that perturbs $\mathbf{d}$ by the minimum amount permitting $\mathbf{d} + \mathbf{r}$'s prediction by the columns of **A**.

With total least squares we assume that in addition to a vector $\mathbf{r}$ perturbing $\mathbf{d}$ there is a matrix of errors **E** perturbing **A**, and in lieu of Eq. (8.22) we have

$$(\mathbf{A} + \mathbf{E}) \cdot \mathbf{X} = \mathbf{d} + \mathbf{r}. \tag{8.23}$$

Equation (8.23) may be written as

$$(\mathbf{A} : \mathbf{d} + \mathbf{E} : \mathbf{r}) \cdot \begin{pmatrix} \mathbf{X} \\ -1 \end{pmatrix} = \mathbf{0}. \tag{8.24}$$

The notation $\mathbf{A} : \mathbf{d}$ refers to the $m \times (n + 1)$ matrix formed by appending $\mathbf{d}$ as an additional column of **A**. A similar remark pertains to $\mathbf{E} : \mathbf{r}$. Instead of minimizing $\|\mathbf{r}\|_2$ subject to Eq. (8.22) with total least squares we minimize $\|\mathbf{E}:\mathbf{r}\|_E$, the Euclidean norm of the matrix $\mathbf{E}:\mathbf{r}$, subject to Eq. (8.23) or Eq. (8.24). Notice the use of the Euclidean norm, defined by Eq. (2.31), with total least squares. Ordinary least squares minimizes the vector norm $\|\mathbf{r}\|_2$. As Chapter 2 explains, for vectors the L_2 and the Euclidean norms coincide. But for matrices the two norms are distinct, and for total least squares, minimizing a matrix norm, we must distinguish between the L_2 and the Euclidean matrix norms.

Golub and Van Loan (1980) show that the total least squares problem corresponds with minimizing

$$\frac{\|\mathbf{r}\|}{(1 + \mathbf{X}^T \cdot \mathbf{X})^{1/2}} = \min. \tag{8.25}$$

Equation (8.25) represents minimizing the perpendicular distance between the data points and the fitting line instead of the vertical distance, as with ordinary least squares. The term "fitting line" is appropriate for two dimensions; for higher dimensions we should refer to a fitting surface or hypersurface. Because we minimize the perpendicular distance of Eq. (8.25) the total least squares problem is sometimes referred to as orthogonal least squares.

Equation (8.25) furnishes one possible way of obtaining a total least squares solution: apply any of the minimization techniques of Chapter 7 to Eq. (8.25) with $\mathbf{r}$ replaced by $\mathbf{A} \cdot \mathbf{X} - \mathbf{d}$. But such a suggestion is hardly sensible because it converts a linear problem into a nonlinear one.

Nor does taking the gradient of Eq. (8.25) and setting it to zero help much. We end up with

$$(1 + \mathbf{X}^T \cdot \mathbf{X})\mathbf{A}^T \cdot (\mathbf{A} \cdot \mathbf{X} - \mathbf{d}) = \mathbf{X}(\mathbf{r}^T \cdot \mathbf{r}). \tag{8.26}$$

Unlike the pleasant situation with ordinary least squares, where the right-hand side of Eq. (8.26) would be zero and convert the equations into a linear system for $\mathbf{X}$, Eq. (8.26) is a nonlinear system of $\mathbf{X}$. We pay a high price for minimizing the perpendicular rather than the vertical distances.

Fortunately, the SVD comes to our rescue. Rather than go into details, which are complicated and whose rationale is far from evident, we shall present the computational procedure in cookbook fashion. The interested reader should consult Golub and Van Loan's article (1980) for the details.

We can prove that the smallest singular value of $\mathbf{A}:\mathbf{d}$ coincides with the minimum Euclidean norm of $\mathbf{E}:\mathbf{r}$. Let s_{n+1} be this smallest singular value. The total least squares solution is given by

$$\mathbf{X} = (\mathbf{A}^T \cdot \mathbf{A} - s_{n+1}^2 \mathbf{I})^{-1} \cdot \mathbf{A}^T \cdot \mathbf{d}. \tag{8.27}$$

Equation (8.27) involves the SVD of $\mathbf{A}:\mathbf{d}$ plus the computation of $\mathbf{A}^T \cdot \mathbf{A}$ and $\mathbf{A}^T \cdot \mathbf{d}$ plus the subsequent matrix inversion, a substantial amount of work. Because we only need s_{n+1}, considerable labor can be saved by calculating s_{n+1}^2 as the smallest eigenvalue of $(\mathbf{A}:\mathbf{d})^T \cdot (\mathbf{A}:\mathbf{d})$ by the power method, with origin shift to give the smallest eigenvalue, of Section 2.3. With proper organization of the programming details we only need accumulate $A^T \cdot A$, $A^T \cdot d$, and $A_j^T \cdot d$, where A_j is the jth column of A, and $d^T \cdot d$; $A_j^T \cdot d$, and $d^T \cdot d$ are needed to calculate s_{n+1}^2. By taking advantage of symmetry, an array of size $(n + 1)(n + 2)/2$, plus some ancillary vector arrays, should suffice.

In some instances the total least squares problem may have no solution. These instances are of more theoretical than practical importance, because the ingeniously contrived circumstances for which they occur are unlikely in practice. We may, nevertheless, check for the total least square's solution not existing. The check is simple. Calculate the smallest singular value of A; call it s_n'. If $s_n' > s_{n+1}$, then the total least squares solution exists and is given by Eq. (8.27); otherwise, it does not.

Rather than calculate an SVD for both A and $A:d$, computationally expensive, Golub and Van Loan (1980) present a recipe that involves only the SVD of A. From the SVD of A we find the ordinary least squares solution via

$$X = V \cdot S^{-1} \cdot U^T \cdot d. \tag{8.28}$$

The truncated version of the SVD, with U^T an $n \times m$ matrix, suffices for Eq. (8.28). Then make use of what Golub and Van Loan (1980) call the total least squares secular equation,

$$s^2 \left(\frac{1}{\lambda^2} + \sum_{i=1}^{n} \frac{c_i^2}{s_i^2 - s^2} \right) = r^T \cdot r \tag{8.29}$$

to find $s = s_{n+1}$. In Eq. (8.29) $r^T \cdot r$ is the sum of the squares of the residuals from the ordinary least squares solution, the s_i's are the singular values of A, and the c_i's are the row components of $U^T \cdot b$; we shall discuss λ shortly, but for the moment set $\lambda = 1$. Then calculate the total least squares solution from

$$X = V \cdot (S^2 - s^2 I)^{-1} \cdot S \cdot U^T \cdot d. \tag{8.30}$$

Notice the similarity between Eqns. (8.30) and (8.28). With neither equation do we have to invert a matrix.

To solve Eq. (8.29) any standard root-finding technique may be used. The interval $[0, s_n]$ brackets the root s, under the supposition that s exists, and Eq. (8.29) is monotonic within the interval. We could, therefore, employ the binary chop, also known as the Bolzano process, as an effective root seeker. This simple procedure is described in nearly any book on numerical methods. Press, Flannery, Teukolsky, and Vetterling (1986, p. 247) give a brief (twenty-one lines of FORTRAN code) function subroutine to implement a binary chop.

But the binary chop is so simple that it can even be written in BASIC. Here are a few lines of BASIC code for finding the root once an interval $[A1, A2]$ containing it has been determined.

```
10   INPUT A1
20   INPUT A2
30   A = A1
40   GOSUB 500
50   F1 = F
60   A = A2
70   GOSUB 500
80   F2 = F
90   A3 = (A1 − A2)/2 + A2
100  A = A3
110  GOSUB 500
120  F3 = F
130  IF SGN (F3) = SGN (F1) GOTO 160
140  A2 = A3
150  GOTO 170
160  A1 = A3
170  IF (A2 − A1) > 0.000001 GOTO 30
180  PRINT A3
```

GOSUB 500 branches to a subroutine to evaluate the function whose root we are seeking. For Eq. (8.29) the function would be the right-hand side of the equation minus the left-hand side. The tolerance in sentence 170 can be changed, but should never be less than the machine epsilon. With each iteration (binary chop) we gain one bit of precision in the root.

If no s satisfies Eq. (8.29), then there is no total least squares solution for the problem. Calculating the root of Eq. (8.29), if one exists, is much less computationally demanding than the complete SVD of the matrix $A : d$.

As the discussion following Eq. (8.27) mentions, we can avoid the SVD altogether by finding the singular values s'_n and s_{n+1} from the power method. Then apply the check $s'_n > s_{n+1}$ and find the total least squares solution, if it exists, from Eq. (8.27). Such a procedure saves labor, but may suffer from accuracy problems if the smallest eigenvalues of $A^T \cdot A$ and $(A : d)^T \cdot (A : d)$ are clustered closely.

To illustrate the method we shall calculate the total least squares solution of Eq. (5.64), used previously with iteratively reweighted least squares and the L_1 method, using the Golub and Van Loan (1980) procedure. The SVD, truncated version, of Eq. (5.64) results in, to five decimals,

$$S = \begin{pmatrix} 7.69121 & 0 \\ 0 & 0.91937 \end{pmatrix}, \quad V = \begin{pmatrix} 0.26693 & -0.96371 \\ 0.96371 & 0.26693 \end{pmatrix},$$

$$U = \begin{pmatrix} 0.16001 & -0.75789 \\ 0.28531 & -0.46755 \\ 0.41061 & -0.17720 \\ 0.53591 & 0.11314 \\ 0.66121 & 0.40349 \end{pmatrix}.$$

From $\mathbf{U}$ and the right-hand side of Eq. (5.64) we obtain, corresponding to Eq. (8.29),

$$s^2\left(1 + \frac{17.61213}{59.15476 - s^2} + \frac{0.18787}{0.84524 - s^2}\right) = 1.2.$$

(Rather than calculate the ordinary least squares solution of Eq. (5.64) we used the results and the residuals already given in Chapter 5.)

The root of the secular equation, computed by the binary chop, is $s = 0.76782$. Because $0.91937 > 0.76782$ the total least squares solution exists. From Eq. (8.30) we find

$$\mathbf{X} = \begin{pmatrix} 0.26693 & -0.96371 \\ 0.96371 & 0.26693 \end{pmatrix} \begin{pmatrix} 59.15476 - 0.58954 & 0 \\ 0 & 0.84524 - 0.58954 \end{pmatrix}^{-1}$$

$$\cdot \begin{pmatrix} 7.69121 & 0 \\ 0 & 0.91937 \end{pmatrix} \begin{pmatrix} 4.19668 \\ -0.43344 \end{pmatrix}$$

or

$$\mathbf{X} = \begin{pmatrix} 1.64903 \\ 0.11514 \end{pmatrix},$$

substantially different from the ordinary least squares solution of $\mathbf{X}^T = (0.6 \ 0.4)$. The residuals from the total least squares solution are, to five decimals,

$$\mathbf{r} = \begin{pmatrix} 0.39553 \\ 0.45513 \\ -0.00288 \\ -0.46089 \\ 0.11631 \end{pmatrix}$$

as calculated from Eq. (8.25). Notice that the sum of the squares of the residuals, $\mathbf{r}^T \cdot \mathbf{r} = 0.58954$, coincides with the smallest singular value of $\mathbf{A} : \mathbf{d}$. Golub and Van Loan (1980) demonstrate that this is true in general.

Figure 8.2 illustrates the ordinary and the total least squares solution of Eq. (5.64). With the former the x-axis is considered error-free. All of the error, concentrated in the y-axis, is representable as the vertical distance between the fitting line and the data point. With total least squares, error also exists along the x-axis, and the residual vectors become skewed to the perpendicular distance between the fitting and the data point. The correlation in errors between the x-axis and the y-axis complicates the fitting procedure, as measured by the relative ease with which we obtain an ordinary least squares solution compared with the more demanding total least squares procedure.

What about the constant λ in Eq. (8.29)? λ is a measure of the coupling between errors in $\mathbf{d}$ and errors in $\mathbf{A}$. $\lambda = 0$ means all the errors are concentrated in $\mathbf{d}$, none $= \mathbf{A}$. For this value of λ, total least squares degenerates to ordinary least squares because s_{n+1} in Eq. (8.27) becomes zero: $s = 0$ is the only solution

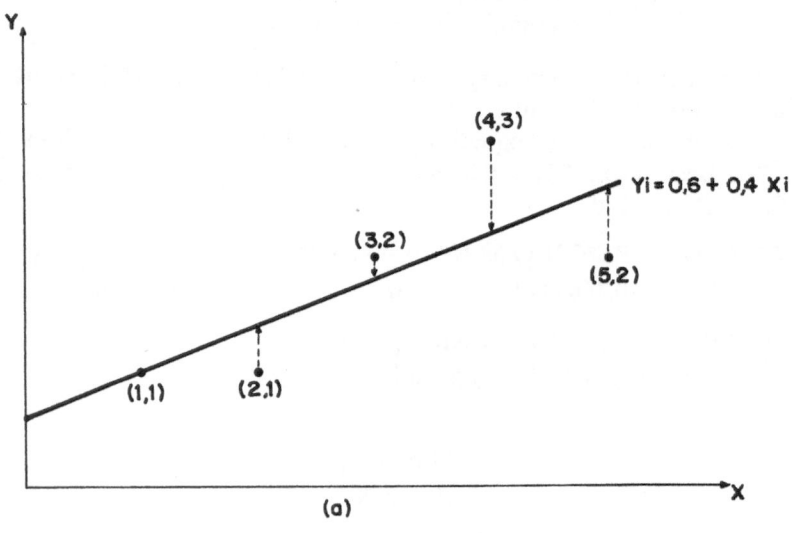

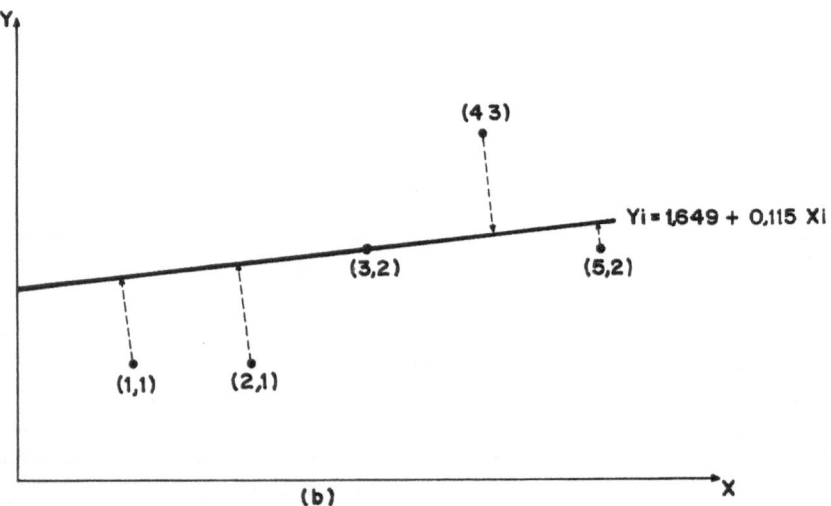

Figure 8.2. Least squares data fitting: (a) ordinary—minimize vertical distances (b) total—minimize perpendicular distances.

of Eq. (8.29). This can be seen by solving the equation as λ assumes larger and larger values. In the limit as $\lambda \to \infty$, $s \to 0$. Alternatively, the smallest eigenvalue of $(A : \lambda d)^T \cdot (A : \lambda d)$ is zero for $\lambda = 0$. $\lambda = 1$ apportions the error equally between d and A, and as λ becomes larger A absorbs more and more of the error. For $\lambda = \infty$ all of the error is concentrated in A. To assess the effect of assumed data errors in A we can solve Eq. (8.27) for different values of λ and calculate the corresponding least squares solutions.

As the discussion following Eq. (8.27) mentions, we can find a total least squares solution without recourse to the SVD. Figure 8.3, adopted from the FORTRAN program in Branham (1989), gives a code to calculate a total least squares solution using normal equations, rather than the SVD. Figure 8.3 parallels closely the program of Figure 5.1 for an ordinary least squares solution. The subroutines DECOMPOSE, SOLVE, and COV accomplish what their in-line counterparts in Figure 5.1 do; decompose the normal equations into their Cholesky factors, solve the normal equations, and calculate the covariance matrix from the inverse Cholesky factor which TRI-INV furnishes.

The smallest singular values s'_n and s_{n+1} are found by the power method of Section 1.5 as the smallest eigenvalues, s'^2_n and s^2_{n+1}, of the normal equations $A^T \cdot A$ and the augmented normal equations $(A : d)^T \cdot (A \cdot d)$. Only the upper symmetric parts of these matrices are stored in a one-dimensional array, and their elements are indexed by the mapping $k = j(j-1)/2 + i$. The normal equations occupy $n(n+1)/2$ memory locations. To form the augmented normal equations it is only necessary to add the n vector $A^T \cdot d$ and the scalar $d^T \cdot d$ to the end of the storage required for the normal equations for a total of $n(n+1)/2 + n + 1 = (n+1)(n+2)/2$ memory locations.

Rather than calculate the smallest eigenvalue by shifting, sometimes subject to slow convergence, the program of Figure 8.3 adopts a different strategy, explained in Branham (1989), based on the observation that if λ is the dominant eigenvalue of A then

$$A \cdot X = \lambda X \quad \Rightarrow \quad A^{-1} \cdot A \cdot X = \lambda A^{-1} \cdot X \quad \Rightarrow \quad A^{-1} \cdot X = \lambda^{-1} X.$$

A^{-1} has eigenvalues reciprocal to those of A. If λ_{min} is the smallest eigenvalue of A, λ_{min}^{-1} will be the dominant eigenvalue of A^{-1} and easily found by the power method, whereas λ_{min} itself may require many iterations to calculate by shifting. The program of Figure 8.3, therefore, applies the power method to the covariance matrix and calculates its dominant eigenvalue, whose reciprocal is the desired smallest eigenvalue of the normal or augmented normal equations.

For most problems the code of Figure 8.3 should be adequate and is considerably faster than use of the SVD. Branham (1989) applied both the SVD approach and the normal equations approach to calculate the total least squares solution of a 20,288 × 41 system. The former consumed ninety minutes of CPU time whereas the latter needed only eighteen minutes. The two solutions agreed to thirteen decimal digits, most likely a consequence of accumulating matrix inner products to high accuracy with cascaded accumulators. (Figure 8.3 uses extended precision rather than cascaded accumulators because the latter are machine-dependent.) There may, however, be sensitive problems for which the greater computational demands of the SVD are justified when calculating a total least squares solution.

In this section we have been concerned with unconstrained, linear total least squares. For nonlinear problems we could, as mentioned previously, minimize Eq. (8.25), using Eq. (7.3) with $p = 2$ for $\|\mathbf{r}\|$, and one of the nonlinear minimization techniques of Chapter 7. Constrained, linear problems may be tackled by the method of Section 5.6, with Eq. (5.92) solved to yield a total least squares solution. The same remark made at the end of Chapter 7 pertains to the constrained, nonlinear total least squares problem: an approximate solution may be obtained if we weight highly the equations of condition representing the constraints.

We may also calculate total solutions in norms other than L_2. For total linear L_1 solutions the reader will find Späth and Watson (1987) informative. To go into details would be too much of a digression. For our present purpose it suffices to mention that the Späth and Watson algorithm relies heavily on the Householder transformation, and obtains the total L_1 solution by an iterative procedure that calculates an ordinary L_1 solution with each pass. The nonlinear, total L_1 problem may be tackled by the simplex method of Section 7.3 applied to Eq. (8.25) with the L_1 vector norm for $\|\mathbf{r}\|$.

8.4. Singular Value Analysis

Return to Eq. (8.28), which arises from the truncated SVD of an $m \times n$ matrix $\mathbf{A}$. Suppose that $\mathbf{A}$ is rank deficient with rank $r < n$. Then $n - r$ of $\mathbf{A}$'s singular values are zero. In general, $\mathbf{d}$ in the system $\mathbf{A} \cdot \mathbf{X} = \mathbf{d}$ lies outside the range of $(\mathbf{A}^T \cdot \mathbf{A})^{-1} \cdot \mathbf{A}^T$, $\mathbf{A}$'s pseudoinverse, and the least squares system

$$\mathbf{X} = (\mathbf{A}^T \cdot \mathbf{A})^{-1} \cdot \mathbf{A}^T \cdot \mathbf{d} \qquad (8.31)$$

has no solution. But if in Eq. (8.28), derivable from Eq. (8.31), we set the reciprocals of the null singular values to zero we obtain a solution $\mathbf{X}$ even though $\mathbf{A}$ is rank deficient. In fact, as long as even one s_i^{-1} is nonzero a full n vector $\mathbf{X}$ will be calculated. But what justification is there for the seemingly bizarre procedure of setting the reciprocals of null singular values to zero?

Consider the residuals from Eq. (8.28). Because orthogonal matrices preserve the Euclidean norm,

$$\|\mathbf{r}\|_2 = \|\mathbf{A} \cdot \mathbf{X} - \mathbf{d}\| = \|\mathbf{U}^T \cdot (\mathbf{A} \cdot \mathbf{X} - \mathbf{d})\| = \|\mathbf{S} \cdot \mathbf{V}^T \cdot \mathbf{X} - \mathbf{U}^T \cdot \mathbf{d}\|. \quad (8.32)$$

In Eq. (8.32) the SVD of $\mathbf{A}$ must be the full, not the truncated, version or $\mathbf{U}^T$ will not be orthogonal. $\mathbf{S}$ is thus an $m \times n$ matrix. Let $\mathbf{y} = \mathbf{V}^T \cdot \mathbf{X}$ and $\mathbf{b} = \mathbf{U}^T \cdot \mathbf{d}$. Then

$$\|\mathbf{r}\| = \|\mathbf{S} \cdot \mathbf{y} - \mathbf{b}\|. \qquad (8.33)$$

$S \cdot y = b$ is a diagonal system with solution

$$y_i = \frac{b_i}{s_i}, \qquad s_i \neq 0, \qquad i < n,$$

$$0 \cdot y_i = b_i, \qquad s_i = 0, \qquad i < n \qquad\qquad (8.34)$$

$$0 = b_i, \qquad i > n.$$

The middle of Eq. (8.34) shows that when $s_i = 0$, y_i can be anything. Whatever its value, the Euclidean norm of r remains the same,

$$\|r\| = \left(\sum_{i=n-r+1}^{m} b_i^2 \right)^{1/2}. \qquad\qquad (8.35)$$

To obtain a unique solution we impose the criterion of smallest Euclidean norm for y, which implies zeroing the y_i corresponding to a null singular value. But from Eq. (8.28)

$$y = V^T \cdot X = S^{-1} \cdot U^T \cdot d, \qquad\qquad (8.36)$$

and zeroing a y_i is accomplished by our setting the corresponding s_i^{-1} to zero.

In Eqns. (8.28) and (8.36) we should be a little careful with the interpretation of S^{-1}. If we use the truncated SVD, S in the equation

$$A \cdot X = U \cdot S \cdot V^T \cdot X = d \qquad\qquad (8.37)$$

is $m \times n$. To go from Eq. (8.37) to Eq. (8.36) we must multiply the left-hand side of the intermediate equation

$$S \cdot V^T \cdot X = U^T \cdot d \qquad\qquad (8.38)$$

by S^{-1}. But because S is $m \times n$, S^{-1} does not, strictly speaking, exist. We should really use the pseudoinverse

$$(S^T \cdot S)^{-1} \cdot S^T,$$

whose properties are defined by Eq. (5.16). The pseudoinverse of the $m \times n$ matrix S is an $n \times m$ matrix whose $n \times n$ diagonal elements are s_i^{-1} and whose last $(m - n) \times n$ columns are null. This is the matrix we mean by S^{-1} in Eqns. (8.28) and (8.36).

To illustrate these ideas let us consider the linear system

$$\begin{pmatrix} 1 & 1 & 6 \\ 1 & 2 & 7 \\ 1 & 3 & 8 \\ 1 & 4 & 9 \\ 1 & 5 & 10 \end{pmatrix} \begin{pmatrix} X_1 \\ X_2 \\ X_3 \end{pmatrix} = \begin{pmatrix} 1 \\ 1 \\ 2 \\ 3 \\ 2 \end{pmatrix}. \qquad\qquad (8.39)$$

The coefficient matrix of Eq. (8.39) is singular—the third column is the second column plus five—and an attempt to use Eq. (8.31) fails. From the SVD of the matrix we calculate

$$S = \begin{pmatrix} 19.65979 & 0 & 0 \\ 0 & 1.86891 & 0 \\ 0 & 0 & 0 \\ 0 & 0 & 0 \\ 0 & 0 & 0 \end{pmatrix},$$

$$V = \begin{pmatrix} 0.11126 & -0.24839 & -0.96225 \\ 0.36717 & 0.91003 & -0.19245 \\ 0.92347 & -0.33190 & 0.19245 \end{pmatrix},$$

and

$$U^T \cdot d = \begin{pmatrix} 4.19983 \\ 0.40178 \\ 0.57562 \\ 0.74139 \\ -0.56479 \end{pmatrix}.$$

Therefore, from the equation $S \cdot y = U^T \cdot d$ we find

$$
\begin{aligned}
y_1 &= 0.21363, \\
y_2 &= 0.21498, \\
0 \cdot y_3 &= 0.57562, \\
0 &= 0.74139, \\
0 &= -0.56479,
\end{aligned}
\tag{8.40}
$$

and from $X = V \cdot y$, with $y_3 = 0$, we get

$$X = \begin{pmatrix} -0.02963 \\ 0.27407 \\ 0.12592 \end{pmatrix}. \tag{8.41}$$

Had we set $y_3 = -1$ instead of $y_3 = 0$ in Eq. (8.40) the solution would be

$$X = \begin{pmatrix} 0.93262 \\ 0.46652 \\ -0.06652 \end{pmatrix}. \tag{8.42}$$

Both of the solutions (8.41) and (8.42) result in residuals whose sum square, 1.20000, is the same and equal to the sum square of the third, fourth, and fifth elements of $U^T \cdot d$. But the Euclidean norm of (8.41) is three times smaller than that of (8.42).

Equation (8.28) affords another way of looking at the SVD approach. If we use the backwards product expansion for the matrix $\mathbf{V} \cdot \mathbf{S}^{-1} \cdot \mathbf{U}^{\mathrm{T}}$,

$$\mathbf{X} = \sum_{i=1}^{n-r} s_i^{-1} \mathbf{v}_i \cdot \mathbf{u}_i^{\mathrm{T}} \cdot \mathbf{d}, \tag{8.43}$$

where, as in Eq. (8.11), $\mathbf{v}_i$ is the ith column of $\mathbf{V}$ and $\mathbf{u}_i$ is the ith column of $\mathbf{U}$ corresponding to the ith reciprocal singular value s_i^{-1}. Equation (8.43) is most useful if we assume that the singular values s_i have been ordered from largest through smallest. The solution is built up from the sum of $n - r$ backwards products, each product of unit rank, weighted by the reciprocal singular value. When the matrix $\mathbf{A}$ is full rank, $r = 0$, and the sum in Eq. (8.43) goes up to n. But when $\mathbf{A}$ is subrank the sum terminates at $n - r$; further terms cannot be added because the s_i's for $i > n - r$ are null. Equation (8.43) clearly shows how we obtain a full n vector $\mathbf{X}$ even though $\mathbf{A}$ may be subrank.

But the discussion so far assumes that a singular value is exactly zero or it is not. Chapter 1 assures us, however, that a singular value is unlikely to be exactly zero. It will more probably be a number of the order of the machine epsilon, or even smaller. To avoid corrupting the sum in Eq. (8.43) by errors that originate from floating-point operations (recall Chapter 1), rather than information supplied from $\mathbf{A}$ and $\mathbf{d}$, we may devise a test for insignificant singular values and discard them. This observation forms the basis for singular value analysis (SVA), a technique of genuine utility for certain problems, but one also subject to abuse.

Rather than check for small singular values close to the machine epsilon, it is more sensible to look at the ratio of the largest to the smallest singular value. The largest singular value may itself conceivably be a relatively small number but as long as the ratio of it to the smallest singular—one definition of the condition number of a matrix—is not too large, the matrix is well-conditioned, and all of the singular values are significant. To be specific, recall from Chapter 2 that the condition number of a matrix is approximately the reciprocal of the machine epsilon—see the discussion following Eq. (2.46). For a well-conditioned matrix, therefore, we want $s_{\max}/s_{\min} < 1/\varepsilon$. But if $s_{\min}$ is much less than $\varepsilon s_{\max}$ the inequality is violated, and the matrix has conditioning problems. Because the small singular values reflect more the computer's round-off or truncation properties than real information, their values are arbitrary. To improve the conditioning of $\mathbf{A}$ we elect to set the small singular values to zero. The net effect is to decrease the rank of the $m \times n$ matrix $\mathbf{A}$ from n to $n - r$, if there are r insignificant singular values, and improve the conditioning of $\mathbf{A}$ by decreasing its condition number of $\mathbf{A}$ from $s_{\max}/s_{\min}$ to $s_{\max}/\tau$, where τ is the cutoff, of the order of $\varepsilon s_{\max}$, below which a singular value is considered insignificant.

Some authorities assert that even other small singular values, larger than the cutoff τ but still substantially smaller than $s_{\max}$, may be set to zero. According to their line of reasoning these small singular values, although no longer merely reflecting the properties of the computer's floating-point arithmetic, are nevertheless corrupted by data errors in **d**, or perhaps by errors in both **A** and **d**, and should also be eliminated. In the author's experience the advice to eliminate small singular values supposedly corrupted by data errors is dangerous. There are, admittedly, situations where the advice is sound. But there are also other situations where zeroing small singular values greater than the cutoff leads to disastrous results.

Typical of the former, where we should eliminate small singular values corrupted by data errors, is digital image restoration. A digital image may be represented by a matrix, where each element of the matrix is a photon count. The image may be manipulated by the techniques of linear algebra, including the SVD. A common problem of image manipulation is the restoration of an image degraded by noise, for example, a blurred image. We decompose the image by the SVD and study the singular values. Those corrupted by noise are discarded and the image reconstructed. When carefully done, the reconstructed image is superior to the original image—blurred images become sharper, for example. Both Pratt (1978, Sec. 14.4) and Andrews and Hunt (1977, Sec. 8.3) discuss the SVD restoration of an image. Unless the singular values associated with noise are eliminated, the reconstructed image, if it can be reconstructed at all, offers no advantage over the original image. Andrews and Hunt (1977, Figs. 8.15–8.18) give examples of the failure to reconstruct an image because of the presence of noisy singular values that should be discarded.

Lack of a strict mathematical relation among the elements of **A** characterizes the SVD approach to image restoration and justifies the elimination of singular values associated with noise. Consider an image of a human face. The elements of **A** are photon counts. Counts within the area of the face will be higher than those for the background. The relation among the elements of **A** is statistical; elements within the face are more highly correlated with one another because of their higher photon count than they are with elements in the background. Elimination of small singular values is also a statistical process, changing the correlations among the elements of **A** to a greater or lesser degree, depending on the suppressed singular values, but violating no strict mathematical relations because there are none to violate. The success of image restoration, the sharpness of a restored from an initially blurred image, for example, depends on criteria that, while quantifiable, are basically subjective: How good does the restored image look?

But the situation with respect to suppressing small but significant singular values changes when the elements of **A** bear a strict mathematical relation to one another. If **A** is the coefficient matrix of equations of condition, **A**'s rows are equalities in an n-dimensional space and represent a certain geometry in that space. Zeroing small singular values does violence to the geometry that leads to the establishment of the linear system in the first place. The author knows of no mathematical justification for setting small but significant singular values to zero and has vehemently criticized such a procedure (Branham, 1979, 1980).

One putative reason encountered in the literature, in Press, Flannery, Teukolsky, and Vetterling (1986, p. 517), for example, for the zeroing of small singular values cannot be sustained. Some claim that by zeroing small singular values we decrease significantly the calculated errors of the solution vector in a least squares solution, but only increase the mean error of unit weight marginally. But this claim is specious.

We can, of course, obtain a least squares solution by the SVD, as Eqns. (8.28), (8.32), and (8.33) show. (Why we would want to do so rather than use computationally less expensive mechanisms, such as normal equations or even orthogonal transformations, will be examined shortly.) However we calculate the solution, Section 5.3 shows that the solution's errors come from the variance–covariance matrix (to the extent that the assumptions of that section are valid). From Eq. (8.31) we deduce that the variance–covariance matrix may be expressed as

$$\mathbf{B}^{-1} = \mathbf{V} \cdot \mathbf{S}^{-2} \cdot \mathbf{V}^{T} \tag{8.44a}$$

or, alternatively,

$$\mathbf{B}^{-1} = \sum_{i=1}^{n} s_i^{-2} \mathbf{v}_i \cdot \mathbf{v}_i^{T}. \tag{8.44b}$$

Although Eq. (8.44a) is derived from substitution of the SVD of $\mathbf{A}$ into the pseudoinverse of $\mathbf{A}$ in Eq. (8.31), it represents the eigenvalue–eigenvector decomposition of $\mathbf{B}^{-1}$: the s_i^{-2}'s are the eigenvalues of $\mathbf{B}^{-1}$ and the $\mathbf{v}_i$'s are the columns of the eigenvector matrix of $\mathbf{B}^{-1}$.

Assume that the eigenvalues in Eq. (8.44) have been ordered from largest to smallest and that their eigenvectors follow the same order. The squared singular values s_i^{-2} are positive as are the diagonal elements of the backwards product $\mathbf{v}_i \cdot \mathbf{v}_i^{T}$. Suppose that r singular values have been set to zero. The summation in Eq. (8.44b) will be truncated before the upper limit n is reached. We then lose the positive contributions from the remaining r terms. If the mean error of unit weight changes only marginally between the full-rank and the subrank solutions, the assumption of the proponents of suppression of small singular values, we underestimate the errors of the solution vector (see Eq. (5.40)). Nor will the correlations obtained from the truncated series represent the actual correlations among the unknowns.

Given that a subrank covariance matrix underestimates the errors of the unknowns, it would be incorrect to compare results of independent researches on the basis of formal mean errors if one research has used a subrank covariance matrix and another the full-rank covariance matrix. The results from the former will obviously look better, but the comparison is also obviously invalid. This is no laughing matter. The author has actually seen it done. See his 1979 paper for a concrete example (Branham, 1979). Some try to conceal the nature of what they are doing when suppressing small singular values by use of turgid terminology, something like "retention of the significant part of the orthogonal sum of two components of the solution vector." But setting small but significant singular values to zero is at best questionable judgment and at worst fraud. Only when the singular value is statistically insignificant, and hence represents more the computer's round-off or truncation properties than genuine information, can some justification be found for zeroing it.

But with proper care we should avoid forming a matrix with a high condition number, which leads to insignificant singular values. The advice given in the discussion at the end of Section 5.4 also pertains to the SVD. We should avoid increasing unnecessarily the condition number of **A** by a poor selection of units.

The scientist tends to view the variables in the model in terms of their usual units, which may not necessarily be the best ones for ensuring good numerical properties for **A**. As a concrete example suppose that we are working with the orbits of minor planets (asteroids). For solar system objects the astronomer selects as the natural units for distance the astronomical unit (AU), very nearly the mean distance of the Earth from the Sun, and for the unit of time the day. But with these units a typical minor planet between Mars and Jupiter will have coordinates of the order of 1 AU and velocities of the order of 10^{-2} AU day^{-1}. The columns of **A** associated with the coordinates will be about 10^2 larger than those associated with the velocities. Whatever conditioning is inherent in **A** will be amplified by 10^2 just by the choice of units.

We should, therefore, see to it that the units are such that the columns of **A** do not vary greatly in magnitude among themselves, either by selecting them to be compatible with good conditioning—in the example of the previous paragraph measuring the velocities in AU per 100 days is more sensible—or, as mentioned in Section 5.4, preconditioning **A** to minimize the variation among columns.

Suppose, as an example, that our interest is a demographic study of population growth during the fourteenth century. We have data for the years 1310, 1320, 1330, 1340, 1350, and 1360. Because population generally increased during the Middle Ages, but declined sharply during the outbreak of the Black Death in the middle of the fourteenth century, we postulate that a parabola, concave downward, fits our data. From the fitting curve

$$y_i = a_0 + a_1 t_i + a_2 t_i^2 \tag{8.45}$$

we find for the coefficient matrix **A**

$$\mathbf{A} = \begin{pmatrix} 1 & 1310 & 1716100 \\ 1 & 1320 & 1742400 \\ 1 & 1330 & 1768900 \\ 1 & 1340 & 1795600 \\ 1 & 1350 & 1822500 \\ 1 & 1360 & 1849600 \end{pmatrix}. \tag{8.46}$$

COND(**A**) is high, $1.3 \cdot 10^{10}$. It would be impossible to tackle this problem in single-precision on most computers; we would have to use double-precision.

Because **A**'s largest singular value is $4.4 \cdot 10^6$ and the smallest $3.4 \cdot 10^{-4}$, an advocate of SVA would be inclined to argue that the smallest singular value is negligible and set it to zero—and get the results that he deserves (probably indicating that population increased during the Black Death).

But the high condition number implies that something is wrong with our matrix. A more sensible approach realizes that the columns of the matrix vary in norm by several orders of magnitude. Column scaling, not SVA, is indicated. Section 5.4 mentions the scaling of each column by its inverse Euclidean norm. Another strategy scales by the power of the base of the floating-point numbers used close to the inverse Euclidean norm. The latter scaling is preferable because it changes only the exponent, not the mantissa, of the floating-point number. Assume that we are using a binary computer. The Euclidean norms of the columns are, to two decimals (great precision is unnecessary): 2.45, 3,270.34, 4,367,684.74. Upon taking logarithms base two of these norms we find, for the scale factors, $2^{-1}, 2^{-12}, 2^{-22}$. When these scale factor are applied to the columns COND(**A**) becomes a much more reasonable $3.6 \cdot 10^4$, sufficiently low to permit single-precision calculations. No singular values need be eliminated because under no hypothesis could the smallest singular value be declared "insignificant".

But an even more sensible approach notices that the variable t_i is deficient. We are interested in a fifty-year period of the fourteenth century, but select a variable whose origin, A.D. 1, is far away. A change to a new variable is indicated, with origin near the middle of our time span and perhaps normalized to a standard interval such as $[-0.5, 0.5]$. With these changes, known as coding and long employed by statisticians, the new variable could be

$$t_i' = \frac{t_i - 1335}{1360 - 1310}. \tag{8.47}$$

With the variable t_i', **A** of (8.46) becomes

$$\mathbf{A'} = \begin{pmatrix} 1 & -0.5 & 0.25 \\ 1 & -0.3 & 0.09 \\ 1 & -0.1 & 0.01 \\ 1 & 0.1 & 0.01 \\ 1 & 0.3 & 0.09 \\ 1 & 0.5 & 0.25 \end{pmatrix}. \tag{8.48}$$

COND(**A'**) is 10, highly satisfactory. We could, of course, use a combination of scaling and coding, hardly necessary for the matrix (8.48). But if we do so and divide the first column of **A'** by two and multiply the last column by four the condition number becomes 2.8. It would be difficult to drive the condition number any lower.

The important observation is that nothing in fitting Eq. (8.45) to our data mandates numerical difficulties. Our problems were caused by a poor choice of variable coupled with failure to scale. SVA is neither needed nor desirable.

Sometimes changing the variable involves operations more complicated than mere coding. Curve fitting with higher-order polynomials than Eq. (8.45), which when done over the interval [0, 1] leads to the notoriouslly ill-conditioned Hilbert matrices, becomes tractable not by coding but by switching to polynomials that are mutually orthogonal over the range, such as Chebyshev polynomials. Forsythe and Moler (1967, Chap. 19) discuss Hilbert matrices. See Fike (1968, Chap. 6), among other references, for a treatment of Chebyshev polynomials.

But suppose that the ill-conditioning is inherent to the problem and cannot be transformed away. Perhaps then SVA may be useful? Hardly. In addition to the theoretical objections to SVA, when used on a matrix derived from a definite mathematical model, already presented, the strictures of orthogonal transformations compared with normal equations mentioned in Section 5.4 are applicable to SVA, perhaps even more so. An inherently ill-conditioned matrix means the time has come to gather more data to strengthen the system, not diddle with singular values. A solution eked out by SVA from poor or incomplete data will be at best a questionable contribution to science and at worst pernicious. (Gresham's law of economics, bad money drives out good, applies to science as well, not always but sufficiently frequently that one wants his results based on the best data possible.) In short, no amount of mathematical technique substitutes for good data.

If more or better data are unobtainable the author believes that it is more honest to state so baldly rather than to present a subrank solution, along with mean errors and correlations, from SVA to which greater importance may be assigned than the solution merits. The saying from computer science, GIGO (garbage in, garbage out), applies equally well to the solution of overdetermined systems.

We finish this section, this chapter, and the book with a few words, not about SVA, but about the SVD and its proper use. Some feel that the SVD, especially when accompanied by SVA, should be the method of choice for attacking least squares problems. The author cannot agree. SVA, unless used for purposes such as digital image restoration, is a dubious procedure, and the SVD is much more computationally demanding than the formation and solution of normal equations; it also requires, for large linear systems, much more computer memory. From the operation counts for the SVD given in Section 8.2, compared with those derived for the normal equations and their Cholesky decomposition given in Section 5.2, plus the $mn^2/2$ operations needed to form the normal equations, we see that the SVD does a minimum of four times more work, with concomitant greater accumulation of round-off or truncation error. The SVD also has its iterative phase, and it seems foolish to replace the finite Cholesky algorithm with an iterative algorithm unless necessary.

Given that the SVD also requires an $m \times n$ array for the matrix $\mathbf{A}$, in which the truncated matrix $\mathbf{U}$ may be accumulated, an n vector for $\mathbf{S}$, and an $n \times n$ array for $\mathbf{V}$ compared with an $n(n + 1)/2$ array for the symmetric normal equations, it becomes difficult to justify its general use over normal equations, particularly when m is large. The linear system, mentioned in Section 6.4, of 21,365 equations of condition with forty-one unknowns, double-precision numbers, would require 7 MB of memory for its SVD and only 6,888 bytes for the formation of the symmetric normal equations and their solution. And while it is true that computer memory keeps expanding in size, it is also true that scientists seem to keep producing programs that fill up available physical memory. Unless a compelling reason can be found for preferring the SVD, its computational and memory profligacy, compared with the more parsimonious normal equations, becomes untenable.

For most problems, where care has been taken to avoid ill-conditioned systems, such a compelling reason cannot be found. The author has encountered in practice only one instance of an overdetermined system unamenable to solution by normal equations, and that one exception owed its extreme ill-conditioning to an improper mathematical formulation. When properly recast the system quickly yielded to the usual normal equation approach.

The proper use of the SVD in the solution of overdetermined systems is restricted to the total least squares problem of Section 8.3 when the coefficient matrix is inherently ill-conditioned and cannot be strengthened by obtaining more data. In this instance, calculating the smallest singular value by the power method may fail because the conditioning of the matrix $(\mathbf{A} : \mathbf{d})^\mathrm{T} \cdot (\mathbf{A} : \mathbf{d})$ may be too high.

For ordinary least squares solutions with inherently ill-conditioned coefficient matrices, the SVD offers no advantage over a reduction of the linear system by orthogonal transformations, although there may be problems for which the calculation of the singular values is useful. The question of calculating a subrank solution from the SVD should never arise, as explained previously.

The advice offered at the end of Section 5.4 with respect to orthogonal transformations also pertains to the SVD. The scientist should certainly include the SVD, like orthogonal transformations, in his mathematical arsenal, but its use should be reserved for special occasions such as inherently ill-conditioned coefficient matrices. For most problems, particularly when care has been taken with the data to avoid extreme ill-conditioning in $\mathbf{A}$ or $\mathbf{A}$ has been preconditioned by scaling, the normal equations should be adequate, and computationally less demanding than the alternatives.

```
C
C
C       PROGRAM FOR TOTAL LEAST SQUARES WHEN ONE FORMS THE NORMAL EQUATIONS IN SYM-
C       METRIC STORAGE MODE. THE EQUATIONS OF CONDITION NEED BE READ IN ONLY ONCE. THE
C       NORMAL EQUATIONS ARE FORMED IN THE ONE-DIMENSIONAL ARRAY A. ELEMENTS OF A ARE
C       REFERENCED VIA THE INDEXING SCHEME K = J * ( J - 1 ) / 2 + I. TO AVOID HAVING TO RE-
C       READ THE EQUATIONS OF CONDITION AN ARRAY AHOLD HOLDS A COPY OF A. THE RIGHT-HAND-
C       SIDES ARE DEVELOPED IN THE ARRAY Y. WE ALSO NEED A COPY OF THIS, YHOLD. INNER PRO-
C       DUCTS ARE CALCULATED USING EXTENDED PRECISION. LAMBDA IS THE COUPLING PARAMETER
C       BETWEEN ERRORS IN THE VECTOR D OF THE RIGHT-HAND-SIDE AND ERRORS IN THE DATA
C       MATRIX A. FOR MOST PROBLEMS LAMBDA = 1 (ERRORS EQUALLY APPORTIONED).
C
C
        PROGRAM TOTAL--LEAST--SQUARES
        IMPLICIT REAL * 8  (A - H , O - Z )
        REAL * 8  LAMBDA
        DIMENSION  A ( 5050 ), AHOLD ( 5050 ), Y ( 100 ), YHOLD ( 100 ), X ( 100 ), CONDEQ ( 101 )
        CHARACTER * 40  FILENAME
        LOGICAL EXIST
        TYPE  * ,  ´ HOW MANY UNKNOWNS ARE THERE ? ´
        ACCEPT  * , N
        TYPE  * ,  ´ WHAT IS LAMBDA ? ´
        ACCEPT  * , LAMBDA
        IF  ( LAMBDA .EQ. 0.0 )  THEN
             EXIST = .FALSE.
        ELSE
             EXIST = .TRUE.
        END  IF
        TYPE  * ,  ´ WHAT IS THE NAME OF THE DATA FILE ? ´
        READ ( * , 5 )  FILENAME
5       FORMAT  ( A40 )
        OPEN  ( 1, FILE = FILENAME, STATUS = ´ OLD ´ )
C
C   INITIALIZE ARRAYS AND CERTAIN VARIABLES.
C
        DO  I = 1 , N
           DO  J = I , N
                A ( J * ( J - 1 ) / 2 + I ) = 0.0
           END  DO
           Y ( I ) = 0.0
        END  DO
        SSQ = 0D0
        M = 0
C
C   READ IN EQUATIONS OF CONDITION AND FORM NORMAL EQUATIONS IN A.
C
1       READ ( 1, 10, END = 15 )  ( CONDEQ ( I ), I = 1 , N + 1 )
10      FORMAT ( D22.15 )
        M = M + 1
        RHS = CONDEQ ( N + 1 )
        DO  I = 1, N
           DO  J = I, N
                A ( J * ( J - 1 ) / 2 + I ) = A ( J * ( J - 1 ) / 2 + I ) + CONDEQ ( I ) * CONDEQ ( J )
           END  DO
           Y ( I ) = Y ( I ) + CONDEQ ( I ) * RHS
        END  DO
        SSQ = SSQ + RHS ** 2
```

Figure 8.3. Program for total least squares. Taken from R.L. Branham, Jr. (May–June 1989), A Program for Total (Orthogonal) Least Squares in Compact Storage Mode, *Computers in Physics*, **3**, with permission from the American Institute of Physics, New York.

```
        GOTO 1
15      TYPE * , ' THERE ARE ' , M , ' EQUATIONS OF CONDITION '
C
C   EARLY EXIT FOR UNDERDETERMINED SYSTEM .
C
        IF ( M .LT. N ) STOP ' SYSTEM IS UNDERDETERMINED '
        REWIND 1
C
C   SET AHOLD EQUAL TO A AND YHOLD TO Y.
C
        DO I = 1 , N
            DO J = I , N
                AHOLD ( J * ( J - 1 ) / 2 + I ) = A ( J * ( J - 1 ) / 2 + I )
            END DO
            YHOLD ( I ) = Y ( I )
        END DO
C
C   DECOMPOSE NORMAL EQUATIONS BY CHOLESKY METHOD AND SOLVE LINEAR SYSTEM.
C
        CALL DECOMPOSE ( A , N )
        CALL SOLVE ( A , Y , X , N )
C
C   AT THIS POINT WE HAVE ORDINARY LEAST SQUARES SOLUTION. WE NEED TO CALCULATE THE
C   MINIMUM EIGENVALUE OF ATRAN * A. THIS IS DONE BY INVERTING THE CHOLESKY FACTOR S
C   OF A = STRANS * S, FINDING AINV, AND ITERATING UNTIL THE MAXIMUM EIGENVALUE IS
C   OBTAINED. ITS RECIPROCAL IS WHAT WE NEED.
C
        CALL TRI-INV ( A , N )
        CALL COV ( A , N )
        CALL MAX-EIGEN ( A , N , SPRIMEMAXSQ )
        SPRIMEMINSQ = 1D0 / SPRIMEMAXSQ
        TYPE * , SPRIMEMINSQ , ' IS SMALLEST EIGENVALUE OF ATRANS * A '
C
C   A HAS BEEN DESTROYED AND MUST BE RESTORED TO FIND MINIMUM EIGENVALUE OF ( A : D )
C   TRANS * ( A : D ).
C
        IF ( EXIST ) THEN
            DO I = 1 , N
                DO J = I , N
                    A ( J * ( J - 1 ) / 2 + I ) = AHOLD ( J *'( J - 1 ) / 2 + I )
                END DO
            END DO
            DO I = 1 , N
                A ( N * ( N + 1 ) / 2 + I ) = LAMBDA * YHOLD ( I )
            END DO
            A ( N * ( N + 1 ) / 2 + N + 1 ) = SSQ * LAMBDA ** 2
C
C   CALCULATE MINIMUM EIGENVALUE OF ( A : D ) TRANS * ( A : D ).
C
            CALL DECOMPOSE ( A, N + 1 )
            CALL TRI-INV ( A , N + 1 )
            CALL COV ( A, N + 1 )
            CALL MAX-EIGEN ( A, N + 1 , SMAXSQ )
            SMINSQ = 1D0 / SMAXSQ
            TYPE * , SMINSQ, ' IS SMALLEST EIGENVALUE OF AUGMENTED MATRIX '
        END IF
```

Figure 8.3 (*continued*)

```
C
C   IF SPRIMEMINSQ > SMINSQ A TOTAL LEAST SQUARES SOLUTION EXISTS AND IS CALCULATED;
C   OTHERWISE TAKE ORDINARY LEAST SQUARES SOLUTION.
C
        TYPE  * , '
        TYPE  * , ' THE ORDINARY LEAST SQUARES SOLUTION : '
        WRITE ( * , 125 )  ( I , X ( I ) , I = 1 , N )
125     FORMAT ( 2X , I2 , 2X , D22.15 )
C
C   TOTAL SOLUTION DOES NOT EXIST.
C
        IF  ( SPRIMEMINSQ .LE. SMINSQ .OR. EXIST .EQ. .FALSE. )  THEN
            TYPE * , ' TOTAL LEAST SQUARES SOLUTION DOES NOT EXIST '
C
C   TOTAL SOLUTION EXISTS - CALCULATE IT.
C
        ELSE
            DO  I = 1 , N
                DO  J = I , N
                    A ( J * ( J - 1 ) / 2 + I ) = AHOLD ( J * ( J - 1 ) / 2 + I )
                    IF  ( J .EQ. I )  A ( J * ( J - 1 ) / 2 + I ) = A ( J * ( J - 1 ) / 2 + I ) - SMINSQ
                END DO
                Y ( I ) = YHOLD ( I )
            END DO
            CALL  DECOMPOSE ( A , N )
            CALL  SOLVE ( A , Y , X , N )
            TYPE  * , ' THE TOTAL LEAST SQUARES SOLUTION :
            WRITE ( * , 125 )  ( I , X ( I ) , I = 1 , N )
        END  IF
        END
C
C
C   SUBROUTINE TO CALCULATE THE COVARIANCE MATRIX IN SPACE OCCUPIED BY THE INVERSE
C   CHOLESKY FACTOR.
C
C
        SUBROUTINE  COV ( A , N )
        IMPLICIT  REAL * 8  ( A - H , O - Z )
        REAL * 16  SUM
        DIMENSION  A ( N * ( N + 1 ) / 2 )
        DO  I = 1 , N
            DO  J = I , N
                SUM = 0.0
                DO  K = J , N
                    SUM = SUM + A ( K * ( K - 1 ) / 2 + I ) * A ( K * ( K - 1 ) / 2 + J )
                END DO
                A ( J * ( J - 1 ) / 2 + I ) = SUM
            END DO
        END  DO
        RETURN
        END
C
C
C   SUBROUTINE TO CALCULATE THE MAXIMUM EIGENVALUE OF A MATRIX BY THE POWER
C   METHOD.
```

Figure 8.3 (*continued*)

```
C
C
      SUBROUTINE MAX–EIGEN ( A , N , SMAX )
      IMPLICIT  REAL * 8  ( A - H , O - Z )
      REAL * 16  SUM , SUMOLD
      DIMENSION  A ( N * ( N + 1 ) / 2 ) , X ( 100 ) , XHOLD ( 100 )
C
C  INITIALIZE EIGENVECTOR COMPONENTS TO UNITY.
C
      DO  I = 1 , N
          X ( I ) = 1.0
      END  DO
      SUMOLD = 0.0
C
C  NORMALIZE EUCLIDEAN NORM OF EIGENVECTOR. NORMALIZATION CONSTANT IS THE
C  EIGENVALUE. EXPONENT OVERFLOW IS UNLIKELY BECAUSE WE NORMALIZE THE EUCLIDEAN
C  NORM WITH EACH ITERATION. HENCE THE TRICK OF EQ. ( 5.66 ) IS UNNECESSARY.
C
5     SUM = 0.0
      DO  I = 1 , N
          SUM = SUM + X ( I ) ** 2
      END  DO
      SUM = SQRT ( SUM )
      SMAX = SUM
C
C  CHECK FOR CONVERGENCE USING RELATIVE ERROR OF EIGENVALUE.
C
      IF  ( ABS ( SUM - SUMOLD ) .LT. ABS ( SUMOLD ) * 1D - 15 )  GOTO 15
      SUMOLD = SUM
      DO  I = 1 , N
          X ( I ) = X ( I ) / SUM
      END  DO
C
C  MULTIPLY EIGENVALUE BY MATRIX .
C
      DO  I = 1 , N
          SUM = 0.0
          DO  K = 1 , N
              IF  ( K .GE. I )  THEN
                  AIJ = A ( K * ( K - 1 ) / 2 + I )
              ELSE
                  AIJ = A ( I * ( I - 1 ) / 2 + K )
              END  ID
              SUM = SUM + AIJ * X ( K )
          END  DO
          XHOLD ( I ) = SUM
      END  DO
C
C  UPDATE APPROXIMATION TO EIGENVALUE.
C
      DO  I = 1 , N
          X ( I ) = XHOLD ( I )
      END  DO
      GOTO 5
15    RETURN
      END
```

Figure 8.3 (*continued*)

References

Acton, F.S. (1970). *Numerical Methods that (Usually) Work* (Harper and Row, New York).

Andrews, H.C. and Hunt, B.R. (1977). *Digital Image Restoration* (Prentice-Hall, Englewood Cliffs, N.J.).

Branham, R.L., Jr. (1979). Least Squares Solution of Ill-Conditioned Systems, *Astron. J.*, **84**, p. 1632.

Branham, R.L., Jr. (1980). Least Squares Solution of Ill-Conditioned Systems, II, *Astron. J.*, **85**, p. 1520.

Branham, R.L., Jr. (1989). A Program for Total (Orthogonal) Least Squares in Compact Storage Mode, *Computers in Physics*, **3**, p. 42.

Fike, C.T. (1968). *Computer Evaluation of Mathematical Functions* (Prentice-Hall, Englewood Cliffs, N.J.).

Forsythe, G., Malcolm, M.A., and Moler, C.B. (1977). *Computer Methods for Mathematical Computations* (Prentice-Hall, Englewood Cliffs, N.J.).

Forsythe, G. and Moler, C.B. (1967). *Computer Solution of Linear Algebraic Systems* (Prentice-Hall, Englewood Cliffs, N.J.).

Golub, G.H. and Reinsch, C. (1970). Singular Value Decomposition and Least Squares Solutions, *Numer. Math.*, **14**, p. 403.

Golub, G.H. and Van Loan, C.F. (1980). An Analysis of the Total Least Squares Problem, *SIAM J. Numer. Anal.*, **17**, p. 883.

Lawson, C.L. and Hanson, R.J. (1974). *Solving Least Squares Problems* (Prentice-Hall, Englewood Cliffs, N.J.).

Mathews, J. and Walker, R.L. (1964). *Mathematical Methods of Physics* (W.A. Benjamin, New York). A second edition of this work was published in 1970. The author has the first edition, referenced in the text.

Pratt, W.K. (1978). *Digital Image Processing* (Wiley, New York).

Press, W.H., Flannery, B.P., Teukolsky, S.A., and Vetterling, W.T. (1986). *Numerical Recipes: The Art of Scientific Computing* (Cambridge University Press, Cambridge).

Späth, H. and Watson, G.A. (1987). On Orthogonal Linear L_1 Approximation, *Numer. Math.*, **51**, p. 531.

Vandergraft, J.S. (1983). *Introduction to Numerical Computations*, 2nd ed. (Academic Press, New York).

Wilkinson, J.H. (1965). *The Algebraic Eigenvalue Problem* (Oxford University Press, Oxford).

Index